Heterogeneous Cyber Physical Systems of Systems

EDITORS

Ioannis Papaefstathiou
Aristotle University of Thessaloniki, Greece

Alkis Hatzopoulos
Aristotle University of Thessaloniki, Greece

Tutorials in Circuits and Systems

For a list of other books in this series, visit www.riverpublishers.com

Series Editors

Amara Amara
IEEE CASS President

Yen-Kuang Chen
VP - Technical Activities, IEEE CASS

Yoshifumi Nishio
VP - Regional Activities and
Membership, IEEE CASS

LONDON AND NEW YORK

Published 2021 by River Publishers
River Publishers
Alsbjergvej 10, 9260 Gistrup, Denmark
www.riverpublishers.com

Distributed exclusively by Routledge
4 Park Square, Milton Park, Abingdon, Oxon OX14 4RN
605 Third Avenue, New York, NY 10158

First published in paperback 2024

Heterogeneous Cyber Physical Systems of Systems/by Ioannis Papaefstathiou, Alkis Hatzopoulos, Amara Amara, Yen-Kuang Chen, Yen-Kuang Chen.

© 2021 River Publishers. All rights reserved. No part of this publication may be reproduced, stored in a retrieval systems, or transmitted in any form or by any means, mechanical, photocopying, recording or otherwise, without prior written permission of the publishers.

Routledge is an imprint of the Taylor & Francis Group, an informa business

Publisher's Note
The publisher has gone to great lengths to ensure the quality of this reprint but points out that some imperfections in the original copies may be apparent.

While every effort is made to provide dependable information, the publisher, authors, and editors cannot be held responsible for any errors or omissions.

ISBN: 978-87-7022-202-0 (hbk)
ISBN: 978-87-7004-312-0 (pbk)
ISBN: 978-1-003-33839-0 (ebk)

DOI: 10.1201/9781003338390

Table of contents

Introduction	7
Meta-Arduino-ing Microcontroller-Based Cyber Physical System Design *by Prof. Apostolos Dollas*	11
Adaptivity and Self-awareness of CPSs and CPSoSs *by Prof. Eduardo De La Torre*	37
Blockchain in Supply Chain Management *by Dr. Harry Manifavas and Dr. Ioannis Karamitsos*	61
Edge Intelligence: Time for the Edge to Grow Up! *by Prof. Theocharis Theocharidis*	95
Putting the Humans in the Middle of the CPS Design Process *by Anna-Maria Velentza*	175
Analog IC Design for Smart Applications in a Smart World *by Prof. Georges Gielen*	211
Powering Cyber-Physical-System nodes by Energy Harvesting *by Dr. Michail Kiziroglou*	235
Hands on Hardware / Software Co-Design *by Dr. Nikolaos Tampouratzis*	253
About the Editors	269
About the Authors	273

Introduction

Cyber-physical systems are the natural extension of the so-called "Internet of Things". They are "systems of collaborating computational elements controlling physical entities"[1]. Cyber Physical Systems of Systems (CPSoS) are considered "The Next Computing Revolution" after Mainframe computing (60's-70's), Desktop computing & Internet (80's-90's) and Ubiquitous computing (00's); this is supported by the fact that most aspects of daily life are rapidly evolving to humans interacting amongst themselves as well as their environment via computational devices (often mobile) while also most advanced systems employ their computational capabilities to interact amongst themselves.

CPSoS enable the physical world to merge with the cyber one. Using sensors, the cyber parts of the systems monitor and collect data from physical processes, like steering of a vehicle, energy consumption or human health functions. The systems are networked making the data globally available. CPSoS make it possible for software applications to directly interact with events in the physical world, for example to measure and react to changes in blood pressure or peaks in energy consumption. Embedded hardware and software systems crucially expand the functionality and competitiveness of vehicles, aircraft, medical equipment, production plants and household appliances. Connecting these systems to a virtual environment of globally networked services and information systems opens completely new areas of innovation and novel business platforms.

Future CPSoS will have many sophisticated, interconnected parts that must instantaneously exchange, parse, and act on detailed data, in a highly coordinated manner. Continued advances in science and engineering will be necessary to enable advances in design and development of these complex systems. Multi- scale, multi-layer, multi-domain, and multi-system integrated infrastructures will require new foundations in system science and engineering. Scientists and engineers with an understanding of otherwise physical systems will need to work in tandem with computer and information scientists to achieve effective, workable designs. In this book, which came out of a seasonal school, basic and advanced issues on the design of the future heterogeneous CPSoS are presented including relevant Blockchain technologies, Tailor-made AI algorithms and approaches, reconfigurable systems, advanced sensor interfaces and human-centered design processes. Certain advanced tools for the design and implementation of the cyber parts of the CPSoS (i.e. FPGA design tools from Xilinx) are also covered.

The first chapter focuses on the evolution of embedded processors, not as much from a historical perspective as from a perspective of enabling technologies and tools. Emphasis is given on the differences between microprocessors and microcontrollers, and how problem

---

[1] https://www.iotone.com/term/cyber-physical-systems-cps/t145

solving with microcontrollers has in the past been performed with in a more resource conscious manner when compared with the development in the present-day platforms such as Arduino. Many examples are given with a main source of these examples being the unmanned space program from NASA's Jet Propulsion Laboratory. The chapter also covers the desired future capabilities of microcontroller development platforms, so that the flexibility afforded to date will be maintained, whereas more efficient solutions will be possible. The main conclusion is that microcontrollers, and especially the commodity items such as the 8-bit microcontrollers, are cheap, and platforms like Arduino have led to the proliferation of new applications by non-expert designers. This huge success brings us to the brink of the need for better tools and platforms which will allow the designer to evaluate alternative design choices, while keeping track of resource utilization. Lacking a better term, we call this need the "meta-Arduino era".

The second chapter is about the benefits of having Cyber-physical systems (CPSs) and Systems-of-Systems (CPSoSs) with the capability of adapting and, even more, self-adapting themselves. Questions like "what is the level of autonomy required by a Martian rover, considering that the control signals from Earth would take more than 4 minutes to reach to Mars, so controlling under these communication latencies would be very hard" are posed as an example for motivating the work. The sentence "Help is only 140 million miles away" from the movie "The Martian" underlines the problem. And, same as for the protagonist of the movie, the system should be reactive, or even proactive, to adapt to changing or even unexpected conditions of operation. There is, therefore, a clear need for adaptation in this kind of systems. The chapter contains some introductory concepts, definitions and classifications, followed by the identified 'adaptation loop', where its components are classified and illustrated with several examples. It is revealed that adaptation is a very beneficial feature of CPSs and CPSoSs in order to have a longer operation life. The ultimate levels of adaptation are coming into play in the self-adaptive systems, which are a must for fully autonomous unattended operations. Self-related properties are desirable in many systems, and they all face common underline issues which are relevant to the adaptation loop (e.g. knowing, measuring, deciding and performing adaptation on a fabric).

Blockchain is a disruptive technology with numerous potential applications in the sustainable sharing economy. The third chapter deals with one of these areas, the Supply Chain Management (SCM), which involves complex CPSoS. It is estimated that the applications of blockchain to global supply chains can result in more than $100 billion in efficiencies. With the usage of blockchain we experience improvements in provenance and traceability in many areas, like food and pharmaceuticals, for example. Provenance is the ability to track any food or pharmaceutics used for consumption, through all the stages, from production to distribution. Traceability is a way of responding to potential risks that can arise to ensure that all food and pharma products are safe for people. Main benefits of using blockchain technology are transparency, efficiency and security between the parties and associated transactions and this chapter highlights how each one of them is applied in SCM CPSoS.

The fourth chapter considers the fact that we live in an era where almost everything is connected, and everyone has several "things" that are considered as edge devices. How do we make these devices smart? How can they "react"? This gives rise to the Edge – Hub (Fog) – Cloud Paradigm. Sensors are performing the data collection of what we are reading in our modern world. They are cheap, can be purchased off-shelf, and can be connected right away on cheap boards, etc. Modern IoT sensors are extremely mobile, very detailed and precise, and most importantly, cheap. Another important driver in emerging cyber-physical systems are the actuators – the actuators can physically manipulate our physical world, and as a result, change the state of our surroundings, impact the environment, and potentially alter our world. But the Sensors/Actuators need a brain to process the data, derive context and put the context into action! We need to shift the computation toward the edge of the network to improve time-to-action and latency, to have greater privacy and security, and to minimize the cost by avoiding costly servers and communication infrastructures. In order to achieve that we need to face the challenges, such as the limited computing resources, by creating algorithms with low complexity, and to minimize the memory and storage requirements in order to run those algorithms on the edge. The main conclusion of the chapter is that Edge Intelligence is here to stay, and we need to embrace it and focus on building energy efficient, cost-effective, robust, and real-time AI algorithms.

Chapter five introduces an interesting perspective stating that the growth of the Cyber Physical Systems (CPS) should be based on more actively incorporating humans within the design loop. The design of CPS will/should consider how humans interact with the physical world, for example how they communicate or move. Humans, and machines (in terms of both hardware and software) must interact and understand each other in order to work together in an effective and robust way. Even in full autonomous scenarios (e.g. fully autonomous cars, robots at factories of the future, etc.), systems are required to explain their behavior to humans, provide them with feedback and allow them to override system actions. Apart from that, all those autonomous scenarios will become plausible only after we have fully understood and model human behavior and human interaction with the physical world. A trust-based relationship must be built between humans and machines so they can work together and achieve effective human-machine integration. Many theories that we are using, or we are about to use in Engineering and Computer Science have been already studied, probably from a different perspective, by Psychologists and Neuroscientists. Studies in Emotion, Brain function and human behavior can/should be used for the design of efficient human-centric CPS; by measuring the emotional state and the cognitive functions we can design, test, calibrate and evaluate a CPS. Cognitive measures such as memory and attention can also give validity and reliability to our studies since we can label our data based on specific scenarios and being sure that a) we indeed measure the effect that we are planning to measure and b) we have the same effect when we repeat the experiment/measurements. Engineers and Psychologists must work together to put the humans in the center of the CPS.

Chapter six presents some challenges for the design of integrated circuits (ICs) for the "smart society", where an increasing number of applications provide intelligent services with unprecedented functionality that will make our daily lives more comfortable. After a short introduction about the ubiquitous presence of sensing in a smart world, the chapter provides guidelines on how to make the sensing circuits small, and how to make the sensing circuits smart. Sensing is used in all applications where the physical world interacts with the electronic world. This happens in almost all application fields, where physical signals are converted into electronic signals, and back. The coupling to electronic processing allows to make many applications smart or intelligent, i.e. that they selectively can adapt and respond to the actual situation. Networked sensing has become ubiquitous in today's smart world, with Internet of Things, personalized medicine, autonomous vehicles, etc. as main application drivers. These applications need many small and networked sensing devices. For the sensor interfaces, this can be realized by adopting highly digital time-based architectures, resulting in technology-scalable circuits with a small chip area and a high power efficiency. Moreover, most applications require extremely high reliability and robustness in the sensing devices against all variations (production tolerances, temperature, supply, EMI, etc.) as well as against degradation over time. As also described in chapter four, modern applications require more and more intelligent computing in the edge; the energy and data bottlenecks can be limited by performing embedded information extraction in the edge (rather than full data), by designing circuits that are fully adaptive in amplitude and time and that preferably have built-in learning capability. High performance at system level can also be created by combining information from multiple "simple" devices, allowing the use of simple, non-sophisticated devices.

The seventh chapter covers the topic of energy harvesting as a local power source, providing information about how to calculate the available environmental energy and to assess the viability of power autonomy for a given application. Energy harvesting is the collection of energy from the environment of a device, for local use. Ten devices per person globally means there is not enough manpower for recharging. Possible solutions include automated recharging, wireless power transfer and energy harvesting. Storage-to-storage power transfer requires maximum efficiency. In contrast, harvesting for a power-flowing source (e.g. sunlight) requires maximum power transfer. These are two different operation points. All the devices in networked CPSoS (or in other words the "things" in the Internet of Things) need power. This power is 5 orders of magnitude more expensive than grid-layer electrical power. Different energy sources for energy harvesting are presented, like thermal, motion, solar, etc. The available power is enough for various duty cycled cyber physical systems. However, customization to each application is required. This is a significant limiting factor / cost-burden. Energy harvesting is a very promising powering method for portable systems. The main challenge is currently its dependence to environmental conditions which limits applicability to bespoke solutions.

Finally, chapter eight presents a hands-on laboratory example for hardware/software co-design.

After completing the design and simulation of a hardware accelerator, it is implemented in real hardware (e.g. a board which includes one Xilinx FPGA MPSoC) to check that the implemented accelerator adheres to the original design specifications when implemented in hardware. For this purpose, designer develops an application (e.g. in C/C++) which is executed in the Processing System – PS (i.e. CPU) of the target MPSoC and calls the accelerator which is implemented in the Programmable Logic - PL (i.e. Reconfigurable resources) of the device. The methodology used to implement a heterogeneous CPSoS in an FPGA MPSoC consists of two steps: a) Implementation of a platform that includes hardware (e.g. a board with a specific FPGA MPSoC), system software (e.g. Operating System (OS)) and system drivers (e.g. drivers for the PS communication with the PL); b) Implementation of the application running on the above platform. The use of the Xilinx Vivado High Level Synthesis (HLS) tool for the implementation and optimization of our application as well as of Xilinx's SDSoC for the interface with the PS is presented. This chapter presents all the necessary design steps namely: (i) Project creation, (ii) S/W-based Simulation, (iii) Design Synthesis, (iv) C/RTL Co-Simulation, (v) Design Optimization and (vi) Interface Declaration. The presented heterogeneous implementation which utilized the reconfigurable resources of the MPSoC is faster by 38 times when compared with the performance of the application when implemented on a low power ARM Cortex-A53 CPU.

Keywords:
Cyber Physical Systems of Systems (CPSoS), Internet of things (IoT), Security, Blockchain, Human Computer Interaction (HCI), smart sensors, energy harvesting, Hardware/software codesign, reconfigurable systems

CHAPTER 01

Meta-Arduino-ing Microcontroller-Based Cyber Physical System Design

Prof. Apostolos Dollas

Technical University of Crete, Greece

1. Presentation — 14
2. Outline — 15
3. JPL Space Probes: issues at hand — 15
4. The Voyager 1 — 16
5. Embedded and Cyber-Physical Systems (Source: Marwedel) — 17
6. Embedded and Cyber-Physical Systems — 18
7. Cyber-Physical Systems for Earth Observation — 19
8. Microprocessors: A Simplified View of the MIPS Microprocessor — 20
9. Microprocessors: a More Realistic Block Diagram of the Intel Itanium IA-64 — 21
10. Microcontrollers: the ATMEL AVR — 22
11. Embedded and Cyber Physical Systems: Microprocessors or Microcontrollers? — 23
12. Flexible vs. Efficient: AVR w/ and w/o Arduino — 24
13. Arduino: A Convenient but Potentially Tricky Abstraction (I) — 25
14. Arduino: A Convenient but Potentially Tricky Abstraction (II) — 26
15. So, should we do Away w/ Arduino? — 26
16. Interrupts and Timers: Give me back that machine — 27
17. A Meta-Arduino Approach (I) — 28
18. A Meta-Arduino Approach (II) — 29
19. Getting to the Meta-Arduino Era — 29
20. More Examples of Microcontrollers in CPS: Watchdog timers — 30
21. It's more complex than it looks: the Mars rover bug (one of all) — 31
22. VLSI/ASIC, GPU, FPGA — 31
23. "Cheap, Fast, Reliable: Choose any two" (plus more such dilemmas) — 32
24. Jupiter, as seen from Voyager 1, Jupiter's Great Red Spot (1979, Source: NASA) — 33
25. A "Star Wars" Perspective — 33
26. Example: Unmanned Aerial Vehicles (UAV) — 34
27. Recap: CPS Systems Everywhere, but Much of their Compute Power goes to Waste — 35

This chapter focuses on the evolution, and even more so, on the near future of microcontroller-based embedded and cyber-physical systems (CPS), not as much from a historical perspective as from a perspective of past and present enabling technologies and tools. Emphasis is given on the differences between microprocessors and microcontrollers, and how system design with microcontrollers entails a different way of thinking vs. system design with microprocessors. In the past, cyber-physical systems have been designed with more resource-efficient but more arcane methods vs. present-day design methodologies. Today, platforms such as Arduino®© offer great flexibility and ease of use but potentially may lead to resource- or power-inefficient designs as the platform is aimed at inexperienced designers and highlights ease-of-use and flexibility. Thus, it is proposed that in the next generation of Arduino-like platforms, notions like cost-performance tradeoffs and global resource allocation are incorporated as a key aspect of the tools, and made accessible to the non-expert designer. Many examples are presented, mostly from the unmanned space program of NASA's Jet Propulsion Laboratory.

1 Presentation

The proliferation of computing resources in everyday life, combined with their low cost in commodity volumes has led to a paradigm shift of how everyday problems are solved. Back in the 1970's, the notion of user input being transformed into electrical signals and a computer controlling the appropriate actuators applied only to high-end systems in which cost was not of primary concern. Indeed, in the 1970's digital computers translating user commands into actuator controls were employed in fighter aircraft such as the F-16. Today, even a modestly priced car will have many if its functions (from its engine and its breaks to passenger comfort features) implemented this way, i.e. user commands processed by computers.

This abundance of computing resources has led to an altogether new way to design systems. A simple automation in the 1970's might have had a so-called "one shot" 555 timer, coupled to an optocoupler and a relay, whereas today the same application may use an Arduino microcontroller and in addition to a pushbutton take commands over WiFi.

The purpose of this chapter is to present many examples of capabilities of "days old" which have more-or-less vanished today. Such capabilities will be presented through many examples, aiming at the end goal of proposing means to regain them through a new generation of development tools. Cyber Physical Systems (CPS) have existed since the 1970's, long before the term CPS or even the more generic "Embedded Systems" (ES) were even coined, let alone more specialized categories such as "Internet of Things" (IoT). This chapter introduces the evolution of Cyber Physical and Embedded Systems, but not in terms of taxonomies, products, and specifications, but in terms of the common traits and characteristics of such systems, which largely remain unchanged for nearly half a century. Although not always the case, the backbone of CPS and ES has been and remain the microcontroller integrated circuits. Microcontrollers usually have a common set of core functions and characteristics (on-chip timer/counters and peripherals, detailed I/O control at the pin level, low cost, etc.). Such characteristics are presented and contrasted to those of microprocessors, although, depending on application, microprocessors and microcontrollers tend to blend into each-other. In more recent years, powerful tools for development, such as the Arduino development tools have offered a substantial boost to CPS and ES development by making microcontroller-based system design easy and accessible through libraries and platforms. At the same time, the generality and flexibility of solutions offered by platforms like Arduino have taken out of the designer's scope some of the detailed resource mapping, which in the past was done manually by experienced designers. This tradeoff between flexibility and optimization calls for a new generation of tools, which in this presentation we will call "meta-Arduino" era. Comparisons will be made to alternative implementation technologies, such as VLSI/ASICs, GPUs, and FPGAs. Interrupts will be shown to be a fundamental "must have" feature of all embedded systems of non-trivial complexity. Many examples and case studies will be presented, together with some basic principles which hold true regardless of implementation technologies, microcontrollers or otherwise. These examples will highlight the intricacy of demanding applications, such as the Mars rover, in order to provide the context for the need to evolve the development platforms to the "meta-Arduino" age.

2 Outline

- From JPL and Deep Space Exploration in the 1970's and 1980's, to the Arduino, and back to JPL...
- What is an Embedded System (and what is a Cyber-Physical System)?
- What is a Microprocessor? What is a Microcontroller?
- Flexible vs. efficient: Design philosophy tradeoffs
- A Meta-Arduino Approach
- Interrupts and Timers: Give me back that machine
- It's more complex than it looks: the Mars rover bug
- Anything Else in Store? VLSI/ASIC, GPU, FPGA
- More Considerations: Learning Even more from the 1970's and 1980's
- Distributed or Centralized? A "Star Wars" Perspective
- Conclusions

3 JPL Space Probes: issues at hand

A most characteristic and successful case of CPS design is the unmanned space program by NASA's Jet Propulsion Laboratory (JPL), which dates back to the 1950's and the Explorer 1 satellite. Given that space probes, some of which went to the outer solar system and beyond, had the most challenging of operating conditions, from cosmic radiation to limited power on-board, solutions had to be found for reliable long-term operation of these systems. The methodology used was so successful, that some of these systems still transmit information well over four decades after their launch, and from locations beyond the edge of the solar system. In this chapter we will see how some problems stemming from the operating conditions were solved, and why many of the system-level capabilities of designers in the 1970's are not available to CPS designers today in the case in which present-day designs are limited only to what is performed automatically by tools and platforms. We note that platforms like Arduino do allow the designer to fully exploit the capabilities of a target system, however, in order to do so the designer will have to step-in the process. We also note that in JPL's space program high-level programming languages were used – it was not all assembly language and low-level hand coding. Thus, to some extent, the issue at hand is a matter of convenience and flexibility vs. efficiency and design optimization, and how future tools can increase efficiency while maintaining flexibility. The rest of the chapter will focus on specific issues, with many examples from the past and the present.

3 JPL Space Probes: issues at hand

- JPL is the unmanned space program of NASA, sending space probes, operating radio telescopes, and much more
- No means to reset manually (operating in outer space)
- Very few resources: Each of the five on-board computers of Voyager 1 on the late 1970's had some 70KBytes of main memory
- How do we deal with cosmic radiation (causing transient errors), memory failures (permanent errors), etc.?
- How do we perform this type of design with microcontrollers?

4 The Voyager 1

JPL's Voyager 1 was launched in 1977 (source: NASA).

In 1990 it took this picture (source: NASA/JPL-Caltech)

In 2019 it still transmits data, but outside the heliosphere – it is in interstellar space

In the picture to the right we see the last image that the Voyager 1 sent to Earth, taken on February 14, 1990. One can notice a single blue pixel close to the middle of the rightmost vertical light brown line. This photograph is known as the "pale blue dot" photograph, and the blue pixel is the Earth itself. Afterwards the camera was shut off, however, at present (late 2019) the Voyager 1 still transmits data to Earth from other sensors, well over four decades after its launch.

The "Pale Blue Dot" image by NASA is in the public domain (source: NASA), the Voyager 1 Entering Interstellar Space artist's concept belongs to NASA, JPL-Caltech (no Copyright indicated).

5 Embedded and Cyber-Physical Systems
(Source: Marwedel)

- Embedded systems (ES) are information processing systems embedded into a larger product (Marwedel)
- Cyber-physical systems (CPS) are engineered systems that are built from and depend upon the synergy of computational and physical components (definition by National Science Foundation).
- Why do we care? Because in general, we use them in a very broad range of applications, from microwave ovens to earth observation systems and from intelligent homes to endangered species monitoring

The main characteristic of embedded systems is the application- or mission-specific nature of the associated computers. Examples of embedded systems are the processors in printers or microwave ovens. Typically, the user does not have access to the computing resources of the system in the sense of being able to program it as a general-purpose computer, although "firmware" updates may be possible to install on the system. In addition to embedded computation, Cyber Physical Systems have physical interfaces to the outside world with which they interact, such as electromechanical components, actuators, and sensors, thus forming a complete system in which computers and mechanical parts operate in synergy. An example of a cyber physical system is a biped robot or a self-driving car.

In more recent times, some of these definitions have to be considered in context, as traditional applications which were a stronghold of embedded systems market (such as mobile phones) have gradually been opened to the experienced user. In this sense, most users of mobile phones do not develop applications (apps), however, there is a large community of app developers outside the vendor of the hardware or the operating system. In subsequent slides we will see in more detail ES and CPS. An excellent textbook on embedded and cyber physical systems is Peter Marwedel's book "Embedded System Design: Embedded Systems Foundations of Cyber-Physical Systems, and the Internet of Things", Springer, 2018 [1] (and its previous editions).

6 Embedded and Cyber-Physical Systems

- Embedded system: a WiFi hotspot or router

- Cyber-physical system (CPS) for Earth Observation: the DART® II tsunami warning system (source: NOAA)

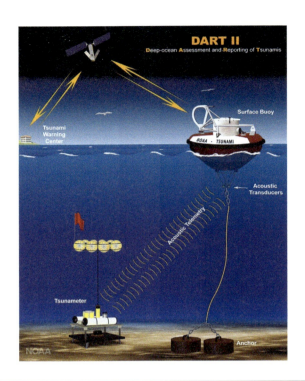

A typical embedded system is a WiFi hotspot or a router. Such a system may not even have a keyboard or a "user interface" in the usual sense of the word, although it does have an interface, be it an Ethernet connector or the (non-tangible but just as real) 802.11 wireless Ethernet protocol. Whether wired or wireless, such a device can be configured, and once it is configured it can provide a number of services to the end user without ever the user having to be concerned about the computing which takes place in the system. More complex and with extensive interaction with physical phenomena in the ocean, computer-to-computer and even surface-to-satellite communication, the NOAA-developed DART®II Deep-ocean Assessment and Reporting of Tsunamis system is not merely a data collection system, but a system interacting with the environment towards timely and accurate detection of tsunamis. Sensors on the ocean floor communicate with buoys, which in turn, as needed transmit relevant information to land via satellite. The real-time operation of such a system is crucial in order to provide sufficient time for population evacuation, whereas its reliability is paramount in order to avoid either false positive or false negative alarms.

DART II is a registered trademark of NOAA, the schematic of DART II appears in the "National Weather Service" site of NOAA (no Copyright indicated).

7 Cyber-Physical Systems for Earth Observation

- Imaging (satellite, drone, and ground based)
- Internet-of-Things: Sensors and CPS inside Bridges to detect fatigue
- Pollution monitoring (land and sea based)
- Air quality monitoring
- All of these systems have microprocessors or microcontrollers
- Many operate under real-time constraints

Going beyond ES and CPS we have the Internet of Things (IoT) in which ES and CPS communicate with each-other and not with a human user or operator. Whereas the typical example of the refrigerator which orders soda pop when it detects that there are only a few cans left may sound somewhat gratuitous, many systems fall in this category, in which large-scale data collection takes place without human intervention. Subsequently, there is a human-in-the-loop as needed, either to post-process data, assess an alarm, or intervene. There exist many cases in point. Many modern buildings have embedded sensors to collect data for a number of parameters to monitor, from structural fatigue in bridges to energy efficiency in large buildings (e.g. hospitals). Pollution is monitored in thousands of stations world-wide with many pieces of data collected each day, but with processing taking place in much larger time windows (e.g. to monitor global warning, or soil erosion). All of these systems have computing resources which are based on microprocessors or microcontrollers, so it is useful to understand the differences between these two classes of computer devices.

8 — Microprocessors: A Simplified View of the MIPS Microprocessor

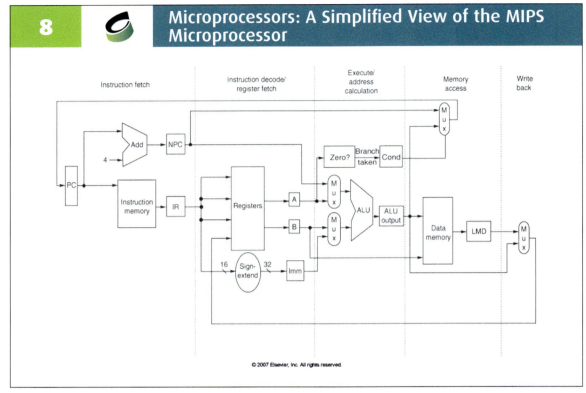

A simplified view of a microprocessor, of the MIPS architecture in this case, shows the main characteristics of all microprocessors. Instructions are fetched from the instruction memory (or from a small, fast memory called "cache", which has selected contents of the main memory), they are decoded (i.e. the control unit checks a field of the fetched instruction to determine what instruction it is), they are executed in one or more steps (e.g. if data –called operands- need to be read from memory or written to it, their address has to be determined first and then the data memory is accessed for the read or write operation) with the use of a unit to perform arithmetic and logic operations (ALU) and the end result of the operation is written in the registers of the processor. At the same time the address of the next instruction is determined, and this cycle is repeated. Although there may be many variants of the above scheme which is specific to the MIPS architecture, e.g. the instruction memory and the data memory may be one and the same, or, in different architectures there may be means to operate on data fetched from memory without having been written in registers first, all microprocessors have the same basic principles of operation, which is to fetch an instruction from memory, decode it, execute it (accessing the data memory if needed), and write back the results. Microcontrollers do have many of the same characteristics, as these integrated circuits also execute instructions coming from some memory, executing these instructions, and writing back the results, but they have major differences as well, which will be presented, below.

Image from "Computer Architecture: A Quantitative Approach" by J. Hennessy and D. Patterson, 4/ed, Elsevier, 2007 © All Rights Reserved

9 — Microprocessors: a More Realistic Block Diagram of the Intel Itanium IA-64

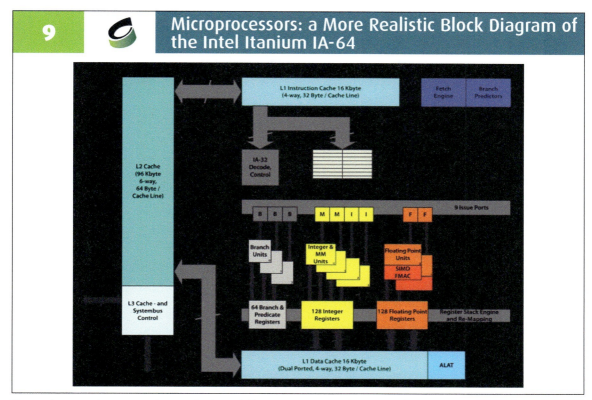

In order to better understand the difference between microprocessors and microcontrollers it is important to understand where they aim at, and what is optimized. A more realistic microprocessor architecture of the Intel Itanium IA-64 is shown, above. Without getting in great depth, resources are devoted (a) to exploit both parallelism (i.e. more than one operations to be performed in every clock cycle) and pipelining (i.e. different parts of the execution of successive instructions to be performed during the same cycle); (b) to ameliorate the "slow" main memory access times - several memories called "cache" memories are on-chip in order for frequently used instructions and data to get to the processor fast (c) to maximize clock frequency, e.g. each operation may be performed in multiple cycles, albeit of a faster clock; (d) to avoid restarting the pipeline as much as possible a "branch unit" aims at the correct prediction of branches which allows the processor to operate more efficiently; (e) to support compute-intensive applications with architecture- and hardware-support of floating point operations. There are many more performance-oriented optimizations, such as support of multi-threaded operation, and in addition, a present-day CPU may have many "cores", each of which is of the above complexity, e.g. eight cores on the same integrated circuit. Such optimizations are common in all microprocessor architectures, and therefore, it is clear that a present-day microprocessor maximizes compute power. This optimization also demands the latest manufacturing technologies (smallest feature size on the integrated circuit), and the largest compute power on a single integrated circuit. Hence, we have maximum compute power, which also comes at a cost (typically in the hundreds of USD or Euros), and associated power requirements.

Image of the Itanium® IA-64 Intel® Corporation, © All rights reserved

10 Microcontrollers: the ATMEL AVR

- A Microcontroller is fully autonomous, including support of several peripheral devices, and hardware support for timers and watchdog timer, but,

- A Microcontroller has substantially less computing power vs. a Microprocessor, and much smaller memory address space (typically 64KBytes vs. at least 4GBytes)

Block Diagram - ATmega32(AVR) - 8-bit Microcontroller

From a first look a microcontroller looks like a microprocessor: it has registers, an arithmetic and logic unit (ALU), and memory to store instructions and data. In this sense, microprocessors and microcontrollers are some forms of Central Processing Units (CPU). However, here is where similarities end. A microcontroller has many necessary peripherals within the integrated circuit, so that it can communicate with the outside world with a serial port (RS 232) or USB, it has a number of timer/counters which can be used to count time (expressed in terms of cycles of the microcontroller's clock) or to count external events, its input/output (I/O) pins can be programmed as needed (e.g. an 8-bit port such as PORT A can have three pins as inputs and five pins as outputs), etc. A microcontroller usually has analog-to-digital (ADC) and/or digital-to-analog converters (DAC), circuitry to detect loss of power, and it generally has features to allow it to reset itself after transient errors, e.g. the so-called "watchdog timer" resets the microcontroller at pre-defined intervals, unless it is re-initialized by software. The operating frequency of a microcontroller is usually low, 4-20MHz (vs. 2-4GHz of a microprocessor), and by comparison to a microprocessor a microcontroller's address space is small (typically 64KBytes), its power requirements are very small, and very importantly it costs very little, with retail prices being a few USD or Euros, and in large volumes it is considered a commodity item. For the remainder of this presentation, when we refer to microcontrollers, it will be of the earlier variety – low cost, modest computing capabilities, with timer/counters, ADC/DAC, etc., such as the ATMEL AVR, although in more recent years there exist microcontrollers with substantial compute power, e.g. for mobile phones.

Image of the AVR® ATmega32 microcontroller® © Copyright 1998-2020 Microchip Technology Inc. Al rights reserved

11. Embedded and Cyber Physical Systems: Microprocessors or Microcontrollers?

- They seem similar but they are different
- Usually the choice is mandated by performance, energy requirements, and other such specifications
- Microcontrollers are everywhere because they are cheap, low power, and reliable, but,
- ...if substantial compute power is needed microprocessors need to be used
- In some cases microprocessors and microcontrollers blend into each-other, e.g. ARM
- How do we develop solutions for CPS with the use of Microcontrollers? It takes a different way of thinking (No Arduinos allowed, yet)

Whereas performance-wise the modest microcontroller is no match for the microprocessor, its "everything on chip" capabilities and low cost lead to a huge market, and especially for many ES and CPS a microcontroller is the technology of choice (e.g. in microwave ovens, garage door openers and other automation applications, wireless phones, etc.). In more recent times the dividing line between microprocessors and microcontrollers has been less well-defined vs. a few years or decades ago, largely because customer needs for more features require ever-increasing compute power. Case in point, mobile phones which for decades were based on microcontrollers, but these days boast powerful multi-core embedded processors to run applications from voice recognition to video streaming. The blend of microprocessors and microcontrollers is taking place in the opposite direction as well, as laptops and other battery-operated general-purpose computing devices require careful energy management and the best degree of integration possible. The use of microcontrollers with very limited resources requires a much different way of thinking vs. solving problems with microprocessors. To illustrate, any C programmer has familiarity with malloc() which gives a program as much memory as it needs, however, with a microcontroller the memory is all physical (not virtual), limited, and for all tasks at hand. Likewise, counting time on a microprocessor may be performed with the help of the operating system, whereas on a microcontroller the appropriate way to do it is with a timer/counter. At present we have not yet examined how the Arduino environment helps the designer of a microcontroller-based system.

12 Flexible vs. Efficient: AVR w/ and w/o Arduino

- Arduino is an environment to make programming of microcontrollers like the AVR easy to use
- Arduino has many libraries
- Arduino has a high-level programming environment, but,
- Arduino does not optimize for good usage of resources in complex systems because it does not "know" what resources are available for each operation.
- E.g. the Arduino will implement a timing loop with code rather than with a timer, unless instructed otherwise.
- So, it is great to use Arduino IF the application is relatively trivial, or IF we know when to step in and hand-code some aspects of the application

Arduino ®, © is an open-source platform for hardware and software designers. Although it is not limited to the ATMEL AVR microcontroller, for historical reasons it was implemented on the ATMEL AVR microcontroller family, and to date enjoys substantial support and a huge user base with this architecture. The philosophy of Arduino was to allow students who were not familiar with electronics or even programming to develop their own applications and see their ideas implemented. This has been accomplished through a series of steps, ranging from an abstraction of the hardware and even the software, to "plug in" modules with peripherals, and an environment which allows it all to work. If a student wants to experiment with a digital thermometer, they only need to buy the temperature sensor, the display, and any of the development systems, and literally in minutes or hours they have their application. Such flexibility has given great boost to the development of prototypes or even products, and the open-source nature of the platform continually gets enriched with new libraries. For example, one can easily build an Arduino-based 3D printer from publically available designs, so not only it is possible to easily develop new applications, but it is even easier to build already-designed systems which suit one's needs. The low cost of the system (one can buy a platform with fairly good capabilities, plus a handful of sensors for less than USD 100), is a tiny fraction of the cost, compared to the electromechanical components of a CPS system. All of the above reasons not only have made Arduino a useful and much loved platform, but they have also led into a very extensive world-wide user base, whose open-source code would be virtually impossible to develop by a single programmer or small corporate team. A downside of this abstraction, however, is that the user does not need to know how microcontroller resources are allocated, and as a result the end design may be wasteful in resources or much less powerful than a design by an expert. Oddly enough, Arduino is an excellent platform for expert designers, because they can take advantage of all libraries and ease of development, and at the same time step-in to allocate system resources, hand code critical sections, and in general take control of the design.

Arduino is Copyrighted and a Registered Trademark of the Arduino Company, Image of Arduino is from the Arduino Company

13 Arduino: A Convenient but Potentially Tricky Abstraction (I)

- Assume that you want to measure time.
- With hand coding we will use TIMERS and the corresponding interrupts, but,
  - Arduino cannot have a global view of how we use resources
  - Arduino will typically implement a timer as a program loop, in which we need only know the number of cycles per instruction (one or two in AVR) and the clock frequency
- Realistic example: Assume that we have a turntable rotating by a stepper motor, with a simple display
  - If timer for stepper motor is implemented with a timing loop the display will go blank during rotation

- **1980's - 2010's : 1 -0 (and counting)**

To illustrate how an inexperienced designer can end up with a poor end result, let us consider a microcontroller which needs to measure time, in order to control a stepper motor. Stepper motors usually need a few dozen steps per second, which may translate to one- or two-hundred events per second. If we have a microcontroller operating at 10MHz (and the AVR typically needs one cycle per instruction), at two hundred events per second, even if we need 50 instructions per event (we typically need a whole lot less) we will need 10,000 instructions, or 0.1% of the compute power of our microcontroller, if we use a TIMER which already exists in hardware. This would require good understanding of timer interrupts and service thereof, as well as allocation of resources – the timer is a physical resource which needs to be allocated for this purpose. With the rest of the compute power of the microcontroller we can run a display and refresh it 60 times per second, and still have plenty of compute power left for more. The Arduino platform does support timers and timer interrupts, however, much of the target audience is not meant to even know what an interrupt is. It is thus quite possible that a timing loop is implemented exactly as such, code which repeats a loop until some pre-determined value, leading to the microcontroller effectively not being able to do anything else. Case in point, a turntable controller which lost its display while the turntable was rotating because the stepper motor controller was implemented with a timing loop (this is an actual design case which came to my attention). Back in the 1980's anyone who would understand the basics of microcontroller operation would also know how to implement such a simple system efficiently. Thus, the challenge is to go to a new generation of tools for the non-expert which allows for efficient solutions of non-trivial problems.

14 Arduino: A Convenient but Potentially Tricky Abstraction (II)

- Assume that you want to build a thermometer for the range of -20C - +70C and accuracy 0.1C
- Is this a fixed point or a floating point application?
- Quite often floating point libraries are used gratuitously, requiring tens or hundreds of instructions where a few instructions would work as well or better

- **1980's - 2010's : 2 - 0 (and counting)**

- **BTW this bad practice appears in pure software as well, e.g. computing of the Mandelbrot set with floating point arithmetic (a notable bad practice)**

A second case example which is also in simplified form a real design which came to my attention, and it refers to a thermometer. If the desired temperature range is -20 to +70 degrees Celsius, and the desired accuracy is 0.1 degrees, a 90 degree range with a resolution of 10 intervals per degree gives us 900 distinct counts, and hence a 10-bit fixed point precision is more than adequate. Even if we choose BCD encoding so that we can drive the display easier (which is actually a good idea), we would need three digits, i.e. 12 bits, plus one bit for the sign. Yet, because of the decimal point, someone may be tempted to use floating point representation, which requires many dozens of instructions for the simplest of operations, let alone conversion to a format suitable for display. In all fairness to Arduino, such bad practices exist even in the pure software domain, and one can find many implementations of the Mandelbrot set (the well known fractals) with floating point, whereas this problem is really best suited for fixed point - there just needs to be some care to place the fractional part of the number correctly. So, is Arduino so problematic? Not necessarily, as we will see.

15 So, should we do Away w/ Arduino?

- First of all, Arduino in the hands of a good designer can do a whole lot more than what it normally does, and,
- What Arduino does not do, a competent designer can develop with hand code (in C or Assembly) and use it together with the rest of the system.
  - e.g. no one wants to write Ethernet or WiFi drivers for scratch, or implement I2C protocols, etc.
- **What do we need then?**
  - Better education and lots of practice in order to take control over the design process, and,
  - New tools to enable us have Arduino-like productivity with 1980's – like design efficiency
  - **We want to Meta-Arduino Cyber Physical System Design, i.e., keep the conveniencies but allow the designer to have more control w/o doing everything by hand**

15 So, should we do Away w/ Arduino?

Any microcontroller designer would prefer to have well tested libraries at their disposal, as well as ease of putting these libraries in a design. A simple WiFi interface would take many person months (if not person years) to implement, but for a few dozen USD or Euros one can have the entire WiFi module for Arduino, including its hardware and software components. Such capabilities are truly enabling at the hands of an experienced designer, who would know how to exploit already existing solutions and libraries, and when to write code in order to complement the existing solutions. We will thus call "meta-Arduino" the notion of better educating the users on good design practices, and at the same time to develop the tools that would enable a designer not merely to have one easy solution but to have several choices and the platform's advise on tradeoffs. If we go back to the timing loop example, the platform could offer implementation of the timing loop with any number of options, whereas at the same time letting the user know of available or taken resources (if TIMER1 is taken, it cannot be used for the stepper motor). It is thus the platform-enabled orchestration of interconnected subsystems and pieces of code which needs to be addressed, which for ES, CPS and IoT applications will lead to an even broader range of developed systems. A more detailed roadmap will be presented, below, once the problem at hand is better understood.

16 Interrupts and Timers: Give me back that machine

- Have you heard the expression "I am Interrupt driven?"
- Main idea for interrupts: stop execution of a program based on an event (external or internal), service the cause of the interrupt, resume execution of the program
- Example: the computer keyboard

Interrupts are important – e.g. we could not have operating systems without interrupts

In order to understand the main reason for the suggestion of the need to evolve tools to the "meta Arduino" era, we need to get into the frame of mind of how problems are generally solved with microcontroller technology. This has to do with "interrupts". Interrupts are the structures which allow for normal operation or program execution to be suspended, a certain code related to the interrupt to be executed, and the program to resume afterwards. Interrupts are caused either by hardware with an internal or external physical signal, or software with program-induced comparable behavior. In microprocessors we have the operating system which based on regular interrupts (e.g. those of the system timer) or as-needed (e.g. the keyboard) manages all processes. In fact, interrupts are the key element

16 Interrupts and Timers: Give me back that machine

in having operating systems altogether. Whereas a microprocessor generally relies on interrupts for such things as I/O, multi-tasking, or response to external events, the usual case is that a program is executed, and when an interrupt is caused (internally or externally) the operating system with run the appropriate code, and schedule tasks according to the desired priorities. By contrast, solving problems with a microcontroller often involves the mapping of processes to associated interrupts. It is thus very common for a timer interrupt to evoke the code for the update of a 7-segment display, whereas a different timer interrupt to sample some external input. The thermometer example which was mentioned earlier is nothing more than an idle loop in the main program, with different on-chip timers controlling input sampling and display. Not only this is a different way of solving problems vs. microprocessor-based design, but even this simple case requires the designer to map what we could call "tiny processes" to the service of on-chip timers. Whereas the Arduino platform clearly supports such design by those in the know, we are far from having the tool which would allow a designer who is not a programmer or electronics/hardware expert to perform such design.

17 A Meta-Arduino Approach (I)

- Microcontrollers have multiple timers/counters (which cause internal interrupts), watchdog timers, interrupts, etc.
- How to solve it - 1: Assume that we want to check for flash floods. Time accuracy of 1 second per observation is sufficient. With a Microcontroller:
  - Put microcontroller to sleep
  - Wake up (from the timer) once per second
  - Check water level
  - If all is well go back to sleep and repeat
    - This way a battery can last for a year, we run 100 instructions per second instead of 4,000,000 (assuming a 4MHz clock frequency)
- Can the above be done w/ Arduino, plus some carefully done hand code? – ABSOLUTELY
- How many 4th year Computer Engineering students can do it, even after having taken an embedded systems / CPS course?

To elaborate further, let us assume that we want to build a system which checks for flash floods. Given that a 1-second sampling is more than adequate, for a proficient designer it makes sense to put the microcontroller to a "sleep mode" and have it wake up once per second. The input is sampled, and if all is well it goes back to sleep, whereas if there is a problem the microcontroller runs all appropriate code for a warning transmission to some base station. Even with a modest 4MHz clock, this means that we only need to run 100 instructions/second vs. 4,000,000 in case we had a continuous sampling loop, which does not seem to matter unless we consider a battery-operated system. The timer-based solution (for the processor alone, because we have peripherals as well) requires 2.5×10^{-5} the amount of energy vs. continuous sampling for the same functionality. There is nothing intrinsic in Arduino which prevents good design practices, except that quite often there is an easy way to make functional, simple projects, which in turn leads to a lack of experience in good design practices. We note that even students taking CPS courses will often opt for the easy way out.

18 A Meta-Arduino Approach (II)

- How to solve it – 2: Assume that we want to measure and send to a remote base station photosynthesis measurements of some plant; log data, possibly take action such as watering. No need to take measurements at night.
- Solution: Keep a real time clock w/ accuracy of 1 sec, keep microcontroller in "sleep" mode, and only take measurements during daytime.
    - The sampling period is different for different times of the day.
    - The clock frequency may be, e.g. 4MHz but the event frequency is 1Hz (or even less).
    - We need to think how often we transmit data in order to optimize energy requirements.
- Can the above be done w/ Arduino, plus some carefully done hand code? – ABSOLUTELY

In a similar case study, if we want to measure the photosynthesis process, log data, and possibly take some action such as watering the plant, we note that there is no need to take measurements at night. We can then take measurements during the day, with a 1sec timer (similar to the previous case study), and in the evening put the system to sleep for the entire night. This way we can have the microcontroller in sleep mode for large periods of time, and use the compute power as needed. Again, the Arduino platform does allow for such designs, however, we may often find designs using the continuous loop approach. However, if there was a means to control all on-chip resources with high level-tools it would be possible to combine ease of use with good design practices.

19 Getting to the Meta-Arduino Era

- The key elements to the above two examples lie in the understanding of the designer of
    - What needs to be done,
    - What can be done well by existing Arduino libraries
    - Where the designer needs to step in.

- There is a broad range of parameters which are inter-related, plus,
- There is a very broad range of ways in which a designer maps a problem to the resources of even a modest microcontroller
- Hence, tools are needed to enable the knowledgeable designer use the resources of the target technology **in a context-sensitive way**

19 Getting to the Meta-Arduino Era

The tools which are needed to get to a meta-Arduino" approach are easy to implement in terms of the platform keeping track of what resources are used by the user so that they will not be oversubscribed, and it is also easy to know if a user's need is met by some library. Not all is easy, however. Such a platform is difficult to implement in terms of how problems are mapped to resources. Even with a handful of resource classes, any designer will employ both time- and space-multiplexing of resources, and the actual solutions for similar problems may be different. If, e.g. we have three classes of events occurring every 11sec, 19sec, and 23sec one possible solution is to use three timers (if we have such available), a second solution is to have one timer with a 1sec resolution (as in some cases there will be events 1sec apart) and a third solution is to know what "phase" each event is in, and determine the value of the timer for the next event. For this example, and assuming all events start at time 0, the successive values of the timer would be 11sec, 8sec, 3sec, 1sec, 10sec, etc. (minus the time duration of the code until the timer is set, if we want to be completely correct). The mapping of problems to resources, however, is the most creative part of the design process. Any experienced designer will have a "bag of tricks" but only a fraction of these tricks are common to many designers. Especially if we consider that design parameters are inter-related and solutions are not universal (in the above example, if the timescale were in msec rather than sec it is possible that the single 1msec timer would work better vs. the more complex solution presented here by reason of the system benefiting less from the use of sleep mode). This poses a real dilemma: how can the tools let the user assess the suitability of an existing library to solve his or her problem vs. writing new code, and keeping the "big picture" in mind?

20 More Examples of Microcontrollers in CPS: Watchdog timers

- Watchdog timers send an interrupt to reset entire microcontroller at a pre-defined interval.
- If the microcontroller operates properly, every so often the watchdog timer is reset and so it never "fires"
- If there is a problem and the system "hangs" the watchdog timer will reset it.

- Remember the JPL space **probes? What** is there to do if there are issues from cosmic radiation? This is how we can re-gain control of a system

As systems become more complex, or they operate in harsh or hostile environments the specifications need to take into account much more than correct functionality in some computational task. Watchdog timers allow systems to be reset if they "hang", as may be the case due to cosmic radiation in a space application. Therefore, a "meta-Arduino" platform would eventually have to evolve into a tool with a holistic approach to design, taking account of correct computation, performance, energy requirements, reliability, and much more...

21 It's more complex than it looks: the Mars rover bug (one of all)

- The Mars rover would crash frequently. What was going on?

- A low-priority process would inherit the priority of a high-priority process, effectively "hogging" the system

- Other high-priority processes would "time out" because they would not have access to the system (bus timeout)

- The time out of high priority processes made the real-time operating system think that the system had crashed (it had not), and so it would reset the entire system, i.e. crash and reset in the process

- Solution: reset (from Earth) one bit in the real-time operating system, regarding process inheritance. Problem solved.

Getting into an example from real life, how does one deal with a problem of a cyber-physical system crashing, apparently randomly, when the physical platform is in Mars? Here we have not only a system with an embedded processor, but also a real-time operating system, multiple sensors, and communication to Earth, all operating in a hostile environment. The problem is detailed in Marwedel's book [1] and other sources, but the solution can be boiled down to the flipping of a single bit in a table of the operating system, which altered the inheritance characteristics of process priorities. The problem was brilliantly detected and fixed from Earth, and the Mars rover was able to proceed with its mission.

Mars Rover Image © NASA. All rights reserved

22 VLSI/ASIC, GPU, FPGA

- The use of microprocessors and microcontrollers is the most flexible way to compute but not always the most efficient. There exist other technologies as well:
    - VLSI/ASIC: Design chips for a specific purpose.
        - + Fastest solution
        - + Best solution for energy consumption
        - - They only do one thing
        - - Due to Moore's law they become obsolete vs. other technologies
    - FPGAs: Map algorithms directly to hardware
        - + Much more energy efficient vs. CPUs
        - + Much faster for specific applications vs. CPUs
        - - Not as fast or energy efficient as ASIC/VLSI
        - - Much design is required
    - GPUs: Not just for graphics anymore
        - + Great for number crunching
        - + Much faster than CPUs
        - - Very demanding in energy vs. FPGAs
        - - Suitable for specific types of problems (mostly the so-called inner loop parallelism)

22 VLSI/ASIC, GPU, FPGA

The scope of this presentation is on ES and CPS and the main technology of concern is that of microcontrollers and microprocessors, as well as the development tools which boost productivity but can lead to non-optimal designs. In this slide we present alternative implementation technologies because they are related, and because either in the form of embedded intellectual property (IP) cores or in the form of co-processors they are widely used in ES and CPS (less so in IoT, so far). The three main technologies to complement microcontrollers and microprocessors are VLSI/ASIC (Very Large Scale Integration/Application-Specific Integrated Circuits), GPU (Graphics Processing Units) and FPGA (Field Programmable Gate Arrays). The examples of their use in ES and CPS are numerous, two such uses are: (a) the inclusion on mobile phones of Bluetooth®, wireless WiFi 802.11, and even video processing IP cores, and (b) the inclusion on computer network routers of VLSI and/or FPGAs in order to perform what is called "stateful packet inspection", i.e. detect network intrusion attacks and other potentially problematic network traffic patterns.

23 "Cheap, Fast, Reliable: Choose any two" (plus more such dilemmas)

- Usually the requirements have conflicting parameters, e.g. high computational demands in battery-operated systems which require maximum autonomy time
- The designer has to really understand the operational parameters – e.g.
    - Q: can you transmit 4K X 4K 24-bit pixel images via a 9600 bps serial port (the JPL space probes had much slower communication channels)?
    - A: yes, but it will take 12 hours.
    - Q: Can we compress the image to reduce its size?
    - A: Yes, but it will require computational time and energy

To make matters more complicated in CPS and ES, even highly optimized designs have to be "good enough" in many different, conflicting parameters – how does a designer compromise between high computational demands (requiring a powerful and power-hungry processor) with long autonomy times in battery-operated systems? Such compromises are necessary in all but the most trivial applications. This poses an additional demand for the "meta-Arduino" types of platforms, as more requirements need to be met. We have thus gone from the present day "easily make something which works" to the desirable "easily make something that works, and uses resources efficiently and without conflict, and allows the designer to understand tradeoff among different ways to map a problem to hardware and software, and allow for custom solutions to be included in the design, and give the designer an assessment on conflicting parameters". One could argue that of the desiderata, above, the "custom solutions" is already partially met, as numerous sensors and IP cores are already implemented in low-cost, easy to include modules (case in point the Arduino module for 802.11), and this is indeed the case today, but such modules are the tip of the iceberg and need to be enriched with more capabilities at the fingertips of the designer. It all often boils down to what is acceptable. The example of the high resolution image transmitted over a slow modem is not just one for the sake of argument…

24 Jupiter, as seen from Voyager 1, Jupiter's Great Red Spot (1979, Source: NASA)

Each image took over 17 hours to reach the Earth

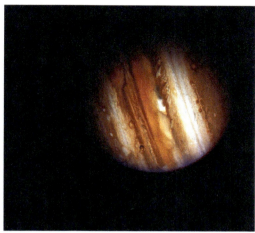

••• but a very real one, as seen in these beautiful photographs taken by Voyager 1 back in 1979. Each image took over 17 hours to reach the Earth, but given the nature of the mission, one such image per day is considered no small feat on its own. For different applications such as a High Quality video TV Box, one would require real-time operation if the user needs to see a sports game, but on a land-based system (wired or wireless) and with ample compute- and electrical power for the appliance. Hence, the question of whether to compress or not, and whether still images or real-time video are desirable is highly context sensitive.

Voyager 1 images © NASA

25 A "Star Wars" Perspective

- In the 1980's, US President Ronald Reagan announced the "Strategic Defense Initiative" (SDI), which became known as Star Wars
- The purpose of SDI was to have an "umbrella" of protection of the US cities against a potential Soviet Union nuclear attack (it was still the cold war)
- Problem: the specifications of the system required the reliability of a decentralized approach, with the efficiency of a centralized approach
- Lesson learned: We need good math to get a solution, and sometimes we can prove that there can be no solution (it is not a matter of getting smart enough people to do it).

25 A "Star Wars" Perspective

Sometimes there exist problems which no platform can address. Such problems may exist from the requirements, the specifications, or even the constraints. The classic example is the "canonical pentahedron", a geometric shape with five identical faces. It is a well-defined object, but one which cannot be constructed (provably). In terms of CPS and ES development, the Strategic Defense Initiative included specifications which were non-implementable, or even contradictory. E.g. it was not possible to implement a system which would have the efficiency of a centralized control to allocate incoming inter-continental ballistic missiles (ICBM) to interception resources, and at the same time have the robustness and reliability of a distributed system. In practical terms, this meant that for the technology of that time, the system would either have to have a single point of failure, or, it would be possible that a computationally distributed solution of incoming ICBMs to geographically separated computers might not lead to a timely allocation of incoming missiles to means for their interception. No matter how good the development platform is, there is no substitute for good mathematical modeling of the problem at hand, with proper dimensioning of the resulting system.

26 Example: Unmanned Aerial Vehicles (UAV)

- Some of the required
- functions:

- Autopilot
- Obstacle Avoidance
- Automatic docking
- Energy management
- Data acquisition: sensors, cameras
- Communication w/ ground station
- Positioning/navigation/path planning
- Robustness and security of communication and navigation
- ...

- Question: One large centralized computer or many communicating independent on-board computers?

Autonomous Earth observation drones are a case in point which encompasses all issues, conflicting parameters, and tradeoffs which were presented here. Their design poses great challenges as they often are battery operated, which means that energy efficiency is important (despite the fact that it is an order of magnitude less than that of the electrical motors in typical cases), with many functions which can be implemented on separate computational units or a more powerful centralized one, and with the desire for high reliability. Although the question in the slide refers to the centralized vs. distributed computing, in reality there are many more, as they include capabilities vs. autonomy, efficiency vs. reliability, incremental design with the need to have provable real-time performance, and efficient but limited architecture vs. an open, extendable architecture which can evolve over time. As Arduino is a platform of choice for the development and proliferation of such drones, this entire presentation could be summarized to the question "How would you like to have platform support in order to not waste time writing low-level software and at the same time have the flexibility to develop a sophisticated drone with well-characterized performance?"

Photo credit: Aristotle University of Thessaloniki, Greece (the HePSOS venue) MPU RX-4 drone with its design team. Photo © ANA-MPA

27 Recap: CPS Systems Everywhere, but Much of their Compute Power goes to Waste

- Microcontrollers are cheap
- Many tools exist for development of new systems
- Many low-cost sensors exist and many more can be developed
- The technology is affordable
- We need to take advantage of existing tools and platforms, including Arduino, but develop next-generation tools which address the inter-related system parameters for efficient CPS design

- **We need better tools and put such tools into the education of the next generation engineers - for lack of better words, let's call such tools Meta-Arduino**

To recap: microcontrollers and especially so the commodity items such as the 8-bit microcontrollers are cheap, and platforms like Arduino have led to the proliferation of new applications by non-expert designers. This huge success brings us to the brink of the need for better tools and platforms which will allow the designer to evaluate alternative design choices, while keeping track of resource utilization. For lack of a better term we will call this need the "meta-Arduino era".

CHAPTER 02

Adaptivity and Self-awareness of CPSs and CPSoSs

Prof. Eduardo de la Torre

Universidad Politécnica de Madrid, Spain

#			#		
1.	Adaptivity and Self-awareness of CPSs and CPSoSs	40	20.	Example #2: Reinforcement learning in ML	52
2.	A use case for motivation: the SPACE environment	40	21.	Options to run Machine Learning	52
			22.	Example #3: Neuroevolution	53
3.	Why adaptation in space?	41	23.	CERBERO CPS Self-Adaptation	53
4.	Basic Concepts	41	24.	CERBERO multi-layer adaptation. CPS and CPSoSs loops	54
5.	Triggers for Adaptation	42			
6.	Types of Adaptation	43	25.	HW Adaptation fabrics addressed in CERBERO	55
7.	Autonomous System Adaptation	43	26.	Monitoring: Unified access to HW & SW Monitors with PAPI	55
8.	Towards more robust and autonomous systems	44			
9.	The adaptation Loop	45	27.	Modelling: Energy vs Execution time vs fault Tolerance	56
10.	Adaptation Loop: Generalities	45			
11.	Adaptation Loop: Monitors	46	28.	Energy vs. Quality of Service via MDC tool	57
12.	Adaptation Loop: KPI Models	46	29.	Self-adaptation loop in a nutshell. CERBERO Options	57
13.	Adaptation Loop: Manager	47			
14.	Adaptation Loop: Adaptation Engine	48	30.	Towards more reliable CPSs	58
15.	Adaptation Loop: Adaptation Fabric	48	31.	Reconfigurable Video Processor	58
16.	Example #1: Evolvable HW	49	32.	Validating the Hardening of MPSoCs for Space Applications	59
17.	System is adaptable and generalizable	50			
18.	Example: Evolvable HW system	50	33.	Conclusions	60
19.	Scalability and evolution for increased fault tolerance	51	34.	Acknowledgements	60

In this chapter we identify the situations where adaptivity is required at both system and system of systems levels for CPSs. Awareness is a preliminary requirement for being aware of the context and about the system itself. These prerequisites open possibilities like self-optimising, self-repairing, self-healing, or self-protecting, among others. However, self-adaptation requires some elements to be present in the system. A formalization of the so-called adaptation loop is presented to show these needs and, as examples, three different adaptive systems are introduced to identify the elements of their adaptation loops: an evolvable hardware system, a machine learning system with reinforcement learing, and a neuroevolvable block-based neural network. Finally, some activities carried out in the CERBERO Project (funded by EU) show some tools available for designing adaptable CPS and CPSoS.

1 Adaptivity and Self-awareness of CPSs and CPSoSs

This lecture is about the benefits of having Cyber-physical systems (CPSs) and Systems-of-Systems (CPSoSs) with the capability of adapting and, even more, self-adapting themselves. This work and some material used in the slides has been achieved within the CERBERO, Enable-S3 and Rebecca projects. I want to acknowledge also the work of my colleagues at the Centre of Industrial Electronics: Andrés Otero, Alfonso Rodriguez, Leonardo Suriano, Arturo Perez, Javier Mora, Rafael Zamacola, Alberto García and Ruben Salvador.

2 A use case for motivation: the SPACE environment

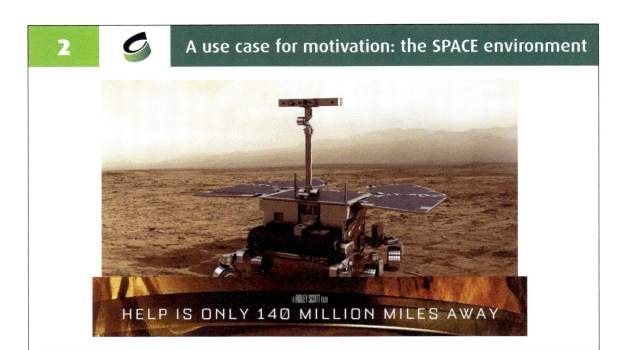

Let's start with an example for motivating the work. What is the level of autonomy required by a martian rover? Take in mind that the control signals from Earth would take more than 4 minutes to reach to Mars, so controlling under these communication latencies would be very hard. I have taken the sentence "Help is only 140 million miles away" from the movie "The Martian" to show such a problem. And, same as for the protagonist of the movie, the system should be reactive, or even proactive, to adapt to changing or even unexpected conditions of operation. There is, therefore, a clear need for adaptation in this kind of systems.

After the motivation, the rest of the lecture will contain some introductory concepts, definitions and classifications, followed by the identified 'adaptation loop', where its components will be classified and will be illustrated with several examples. Then, the approach for adaptation at CPS and CPSoS taken in the CERBERO project will be shown. After this, some details on the work done within he Enable-S3 for hardening SRAM-based FPGAs and the validation platform that was built, are shown. In this section, ultimate adaptation with the use of techniques such as evolvable HW, and its benefits, will be also shown

3 Why adaptation in space?

- Cost is becoming as determinant as radiation
- Resources are starting to be shared
  - Same cameras used for navigation and scientific missions
  - Algorithms change from one function to other → Reconfiguration needed
  - Flash memories for space cannot be rewritten that often
  - SRAM-based FPGAs with many design protections are being considered
- New business models are being projected
  - Farms of satellites offering specific services (Power, communications)
- But, out of LEO, MEO or GEO, systems are fully alone
  - Nobody knows what will be there
  - Some missions are commanded to define Science on the spot
  - Autonomy (Self-*) is required

In the space domain, but also in similar critical domains, there is a new push for reducing costs with the arrival of competitors in the business. The need for cost reduction forces to reuse components, and even the scientific mission and the satellite control part, which were fully separated elements some time ago, now share resources due to this need. For instance, a scientific mission camera could also be used to help in navigation control.

In this context, FPGAs with SRAM technology in their configuration memory are becoming interesting because, although not well hardened against space conditions, offer better cost and more computing performance possibilities. So, apart from missions that orbit Earth that are relatively reachable compared to longer distance missions, poses the requirement of self-adaptation and all type of the so-called self-* features such as self-awareness, self-reconfiguration, or self-healing, among others.

4 Basic Concepts

 Adaptation: *runtime* action *changing structure, functionality and/or parameters of a system*, according to environment, user or self-sensing info.

[F.D. Macías-Escrivá, et al. *"Self-adaptive systems: A survey of current approaches, research challenges and applications"* In Expert Systems with Applications, 2013]

The interpretation by a reconfiguration-oriented engineer

System self-adaptation: combination of *awareness* and *reconfiguration*. Reconfiguration decided *inside the system* itself by a *self-adaptation manager*, which has some degrees of freedom when deciding which modifications to apply. Hierarchical double adaptation is required, where adaptation is set at CPSoS level and at each individual CPS level.

4 Basic Concepts

Adaptation can be considered as a runtime action, i.e., during system lifetime, where the system may be modified to adapt the structure, some parameters or the functionality of the system, and this decision may be taken by external users, third elements, changes in the environment or changes in the system itself, such as the occurrence of faults.

From the point of view of an engineer, system self-adaptation is the combination of awareness and reconfiguration, which is decided inside the system itself by a self-adaptation manager, which has some degrees of freedom when deciding which modifications to apply. Hierarchical multiple adaptation is also required, because adaptation can be set at CPSoS level as well as in each individual CPS level.

5 Triggers for Adaptation

ENVIRONMENTAL AWARENESS: Influence of the environment on the system, i.e. daylight vs. nocturnal, radiation level changes, etc.
Sensors are needed to interact with the environment and capture conditions variations.

USER/EXTERNALLY-COMMANDED: System-User interaction, i.e. user preferences, commands from SoS managers (the boss), etc.
Proper human-machine interfaces are needed to enable interaction and capture commands.
CPSoS decisions are sent to individual CPSs for overall adaptation

SELF-AWARENESS: The internal status of the system varies while operating and may lead to reconfiguration needs, i.e. chip temperature variation, low battery.
Status monitors are needed to capture the status of the system.

The need for adaptation has, in general, three possible triggers:
- By perceiving changes in the environment that make a change recommendable.
- By perceiving changes in the system itself, such as shortage of energy, or presence of faults.
- Commanded by external users or, in the case of CPS that are in a CPSoS environment, by upper-layer decision-taking elements.

6. Types of Adaptation

FUNCTIONALITY-ORIENTED:
To adapt functionality because the CPS mission changes, or the data being processed changes and adaptation is required.
It may be parametric (a constant changes) or fully functional (algorithm changes)

EXTRA-FUNCTIONAL REQUIREMENTS-ORIENTED:
Functionality is fixed, but system requires adaptation to accommodate to changing requirements, i.e. execution time or energy consumption.

REPAIR-ORIENTED:
For safety and reliability purposes, adaptation may be used in case of faults. Adaptation may add self-healing or self-repair features. e.g.: HW task migration for permanent faults, or scrubbing (continuous fault verification) and repair.

Also, adaptation may be classified into three categories according to the purpose for such adaptation.
- It may be functionality oriented, i.e., the system changes its functionality to adapt to other objectives.
- Or extra-functional, where the system keeps doing the same functionality but improving some indicators like performance, energy consumption, quality of service, latency or similar.
- Finally, adaptation may be repair-oriented. The system may be reconfigured so that the operation (also called the performability) of the system can be evaluated and, in cases, faults (transient or repairable ones) may be corrected by adaptation/reconfiguration

7. Autonomous System Adaptation

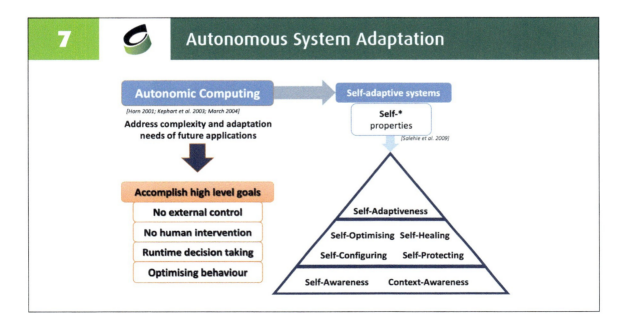

7. Autonomous System Adaptation

Let's review in a couple of slides some definitions of autonomous systems. In 2003 and 2004, the concept of autonomy in computer systems was referred as autonomic computing, where some characteristics were identified such as the ability of accomplishing complex tasks with no external control, no human intervention and being able to do this at execution time with the purpose of optimizing some feature or improve/change the functionality.

The concept of the self-* properties was defined as a pyramid where basic technologies enabled building up newer more advanced techniques. So, self-awareness (being aware of your own status) and context-awareness (be aware of what is around), were considered as the basis of that pyramid. These properties allow to self-configure (adapt), self-protect (improve safety and security autonomously), Self-optimize (some functional or not functional parameter) or self-repair. The vertex of the pyramid is self-adaptation, where the system is able to perform all these properties in a fully autonomous way.

8. Towards more robust and autonomous systems

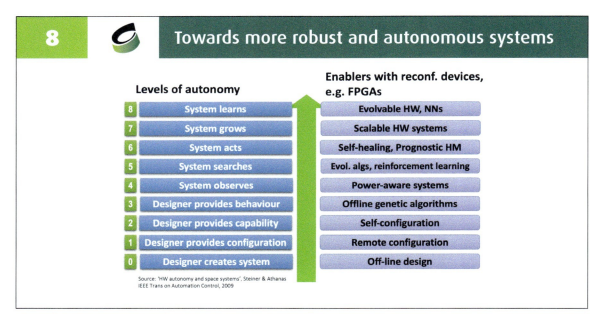

Steiner and Athanas made another classification for space systems, given the fact that the requirements for autonomy are harder than usual (remember the example of the martian rover?). In the bottom layers, the designer provides the capabilities, while in the upper ones, the system itself is able to produce changes. The ultimate levels are systems that can observe, take the decision to change in a sensible way, showing some process of maturity in the execution and even being able to learn from others or from themselves, and consequently adapt. The right hand-side contains some specific enablers of these techniques in the context of reconfigurable computing. We will take a look at some of them along the rest of the lesson.

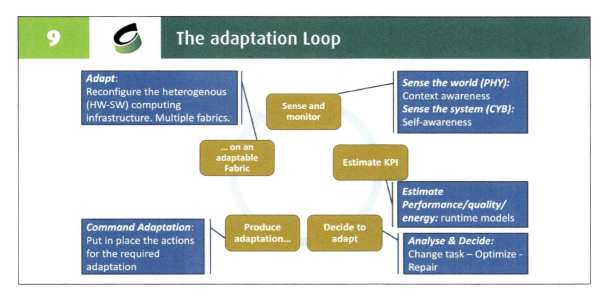

In the second part of this lesson, we will talk about the adaptation loop. It is a generalization of what functions and elements are required to autonomously achieve adaptation in the context of cyber-physical systems. If we start as for the self-* pyramid, we should say that it is required some sensing and monitoring of both internal data (self-awareness or awareness of the CYB part) and the surrounding (awareness of the PHY part or context awareness).

These measured data might not directly be easy to interpret not directly be a decision-taking indicator. So, models that allow to obtain key indicators such as performance, latency, energy or power consumption, reliability and resistance to faults, quality of the service provided, etc., can be obtained via some models.

With the info coming out from these models, a decision taking module should be the trigger for adaptation. Adaptation itself would be done by a reconfiguration engine, able to change the properties of the adaptation fabric..

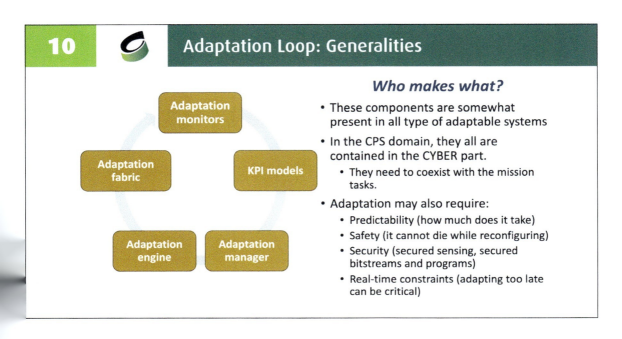

HETEROGENEOUS CYBER PHYSICAL SYSTEMS OF SYSTEMS

10 Adaptation Loop: Generalities

The before mentioned functions are performed by five different modules. These modules do not need to be necessarily explicitly identified in the system, although having their functions clearly identified in a generic manner is beneficial for interoperability between systems.

11 Adaptation Loop: Monitors

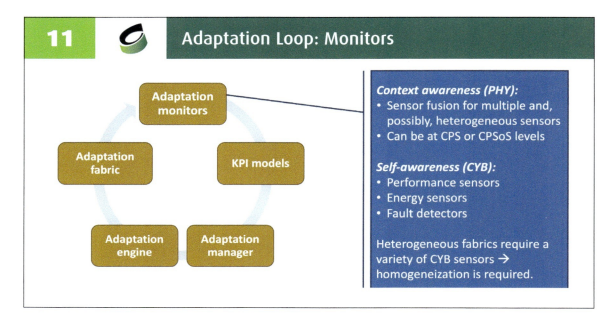

Awareness is composed of two main factors: context awareness, dealing with sensing the surroundings of the CPS, via a combination of sensors of different types, and self-awareness, where the sensors are geared towards offering information of the CPS itself, such as performance monitors, power/energy sensors, fault detectors or similar.

When addressing heterogeneous execution platforms, having a consistent way of accessing all types of fabrics to measure these types of characteristics is a good idea.

12 Adaptation Loop: KPI Models

The information from the sensors is not necessarily directly understandable to take a change decision. In cases, models to produce essential information must be used for deciding adaptation. These models should be lightweight enough to be embedded in the CPS (if self-adaptation is desired), so they should periodically be executed in order to produce the required key indicators that will guide the final decision taking.

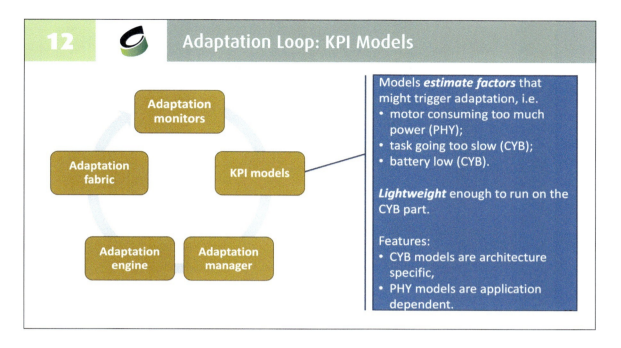

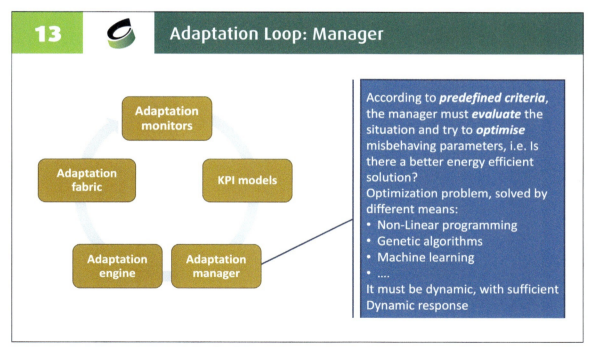

Decision taking is performed by the adaptation manager. After collecting the data, it must follow some rules that ensure that the system Will be able to adapt to optimized situations, either changing to modes of operation that improve some key indicators or changing sme operational parameters. This decision taking, in many cases, means solving some optimization problem, and so, solvers such as non-linear programming solvers, genetic algorithms or machine learning engines can be used. This is accompanied by rule based decision taking, by "intelligence" such as: "If event happens, then set operation mode A or B or C". It can be proactive or reactive.

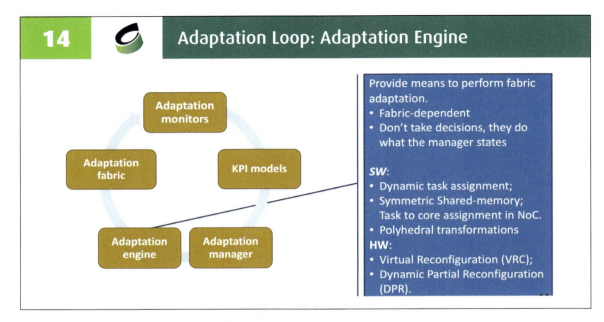

Adaptation manager behaves like 'the boss', but the execution of the adaptation/reconfiguration is achieved by the adaptation engine. It collects the order from the manager and performs the adaptation. If the CPS is heterogeneous, then adaptation is performed by several adaptation engines. Dynamic and Partial Reconfiguration engines are an example to achieve reconfigurability in FPGA fabrics. Priority setting in an OS or remapping of tasks within a multicore platform are examples of SW adaptation.

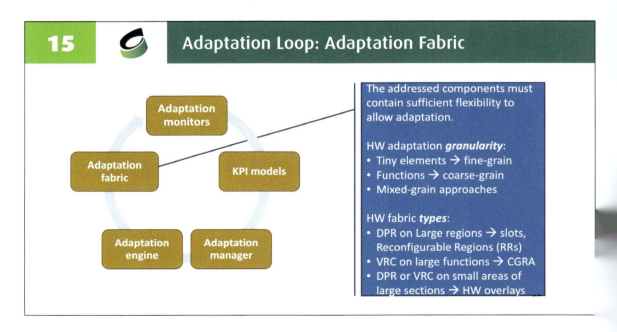

15 — Adaptation Loop: Adaptation Fabric

Of course, in order to finally achieve adaptation, something must be adaptable. Fabrics should have the capability of being changeable. HW fabrics require reconfigurability, and so, FPGAs with SRAM configuration memory are perfect candidates for this, although virtual-reconfiguration is also possible (functions are selected via MUXes). Scalability (being able to extend the work a variable number of elements), diversity (having, for instance, several versions of the same algorithm) into different fabrics, dynamic redundancy (being able to switch on/off components to achieve redundant execution for higher reliability) are examples of properties that contribute to high system adaptivity.

16 — Example #1: Evolvable HW

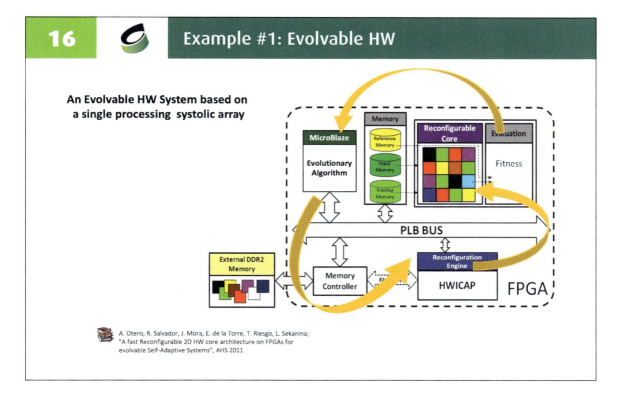

An Evolvable HW System based on a single processing systolic array

A. Otero, R. Salvador, J. Mora, E. de la Torre, T. Riesgo, L. Sekanina; "A fast Reconfigurable 2D HW core architecture on FPGAs for evolvable Self-Adaptive Systems", AHS 2011

As a first example of adaptive system, let us consider evolvable HW (EH). It is a concept where HW may be adapted according to some optimization criteria carried out by an evolutionary algorithm. In this case, the adaptable HW is a systolic array where each element in the array is a processing element taking information from the left and top neighbors and sending the result to the right and bottom ones. Input data gets pixels from a window in an image, same as in a convolutional filter, from the top and left edges of the array. The pixel that is selected in every row/column is also decided by the evolutionary algorithm (EA) The EA proposes a solution, mutates it into several possible candidates, and the one which provides better results with respect to the desired output is selected as the candidate to continue the evolution, following an evolutionary loop that, as for living beings, makes the species (the circuit) to evolve.

17 — System is adaptable and generalizable

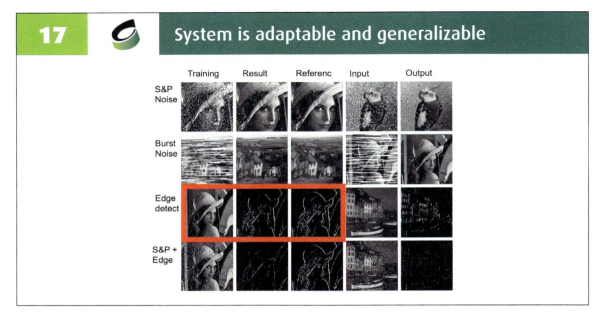

The system works with images, and it may perform different image operations. The system is ONLY fed with the input image and the expected result. The closer the image coming out of the array to the expected result, the better fitness value it will have (fitness is the quantitative value that measures the quality of the resulting circuit). The system has shown to be generalizable (it adapts to different problems, such as different types of noise, edge detection or similar). Once trained with an image, the obtained filter should also work with other images (two rightmost images).

The ege detection feature is highlighted in red, as an example. The reference may be computed offline with an algorithm for edge detection but, after this, by providing the original input and the computed reference, the system is able to obtain a HW system that, much much faster than SW, may do the same task (at over 450 Msamples/second in our case).

18 — Example: Evolvable HW system

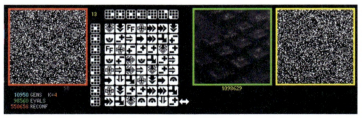

Adaptation Loop:
- **Adaptation Fabric:** Systolic array overlay
- **Monitor:** Fitness compute unit
- **KPI:** Sum of absolute differences (to minimise)
- **Adaptation Manager:** Genetic algorithm
- **Adaptation engine:** DPR on FPGA frames

Results
- Fast evolution: > 140.000 evals/sec, total: 1 sec
- Array works at 450 Mpixels/sec
- Small: 2 CLBs per PE
- Generalizable (noise filtering, edge detection, image enhancement, etc.)
- Scalable (grows or shrinks)
- Self-healing

18 Example: Evolvable HW system

The system also proved to work with noisy inputs and reference, assuming that the noise is not correlated between both. The figure shows how, with a very noisy input image (left) and a very noisy reference (right), the system is capable to produce the output image (the keyboard image) being agnostic of the type of noise and with no equations.

Where are the elements of the adaptation loop in this case? The adaptation fabric is the systolic array, which may be changed by partial reconfiguration (the adaptation engine). The fitness function is the KPI parameter that is obtained with a HW module that performs the summation of the pixel differences between result and reference for all image pixels. In this case, the adaptation manager is the evolutionary algorithm. Since all elements are inside the system, it may be considered autonomous. The PeE in the array were designed so that reconfiguration may be done very fast. A set of parallel arrays in an FPGA achieved a world record of over 145,000 evolutions per second, comprising reconfiguration and evaluation, and so, a couple of seconds is enough for obtaining a good quality noise-agonostic image filter, able to operate at high speeds due to the modular design of the reconfigurable fabric.

19 Scalability and evolution for increased fault tolerance

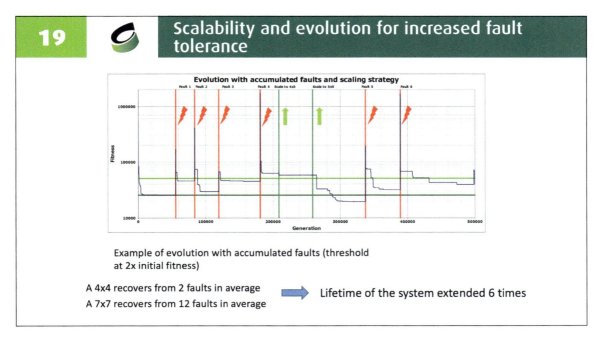

Example of evolution with accumulated faults (threshold at 2x initial fitness)

A 4x4 recovers from 2 faults in average
A 7x7 recovers from 12 faults in average ⟹ Lifetime of the system extended 6 times

Systems based on evolvable hardware, when evolved intrinsically (i.e., the system is evolved/trained in the same fabric that will be used later for normal operation) are capable of self-healing. Evolving in an array with faults in the array makes solutions to bypass the faulty area, getting to newer solutions that are able to solve the same problem. Combining this with scalability, i.e., letting the array grow in dimensions to allocate more PEs may be used for lifelong operation of the circuit. In the image, the system evolves to a low fitness point and, from there, a fitness threshold slightly over it may set the criteria to scale the array up. In the example figure, After first evolution, a second one after a fault injection falls behind the threshold, and as well for the second and third accumulated faults. With the fourth one, the system is not able to recover, and it is scaled twice. With the new size, the system is even able to reach fitness values better that the original one.

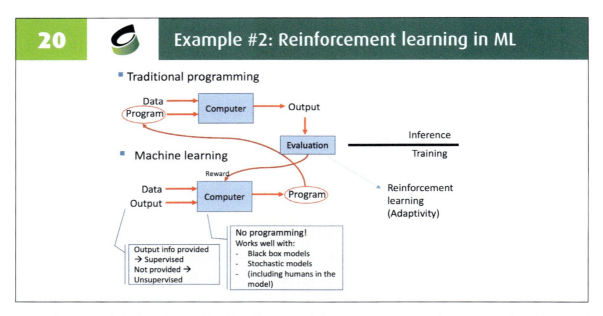

Another example in the field of artificial intelligence can be mentioned. While in conventional programming, programs are designed to transform some input data into some output results, the ML approach uses input data and expected output to produce the program that will, later on, do the inference. At runtime, the process of evaluation of the quality of the results and feeding it back to the training phase to improve the inference method is also a loop where adaptation is used to improve the result. This method is called reinforcement learning.

Options to run Machine Learning

- **Training** in the **cloud** (clusters)
  - Specialised computing services. Essential when mixed with big data and data analytics.
  - Custom tools and frameworks for users (Café, Tensorflow, OpenVINO…)
- **Training** in the **edge**
  - Adaptivity in the context of IoT and autonomous devices
  - Chips for accelerating in embedded (70€!)
- **Inference** in the **cloud**
  - The final result is a service delivered via Internet to people
- **Inference** in the **edge**
  - Cyber Physical systems with fast dynamic response, expert sensors, autonomous systems, adaptative systems, self-aware reactive systems
    - 5G will reduce latency
    - Silicon neuro-chips will provide more computing power
    - Newer technologies will provide energy reduction
    - AI will provide the 'magic' of the service

Opportunities for working cooperatively at edge (CPS) and fog (CPSoS) levels

The training and inference stages may be run in different platforms, but in the context of IoT, it is of special interest doing this in the edge (devices close to the sensors), or cooperatively, close to the edge (referred as fog computing). The specific requirements of low latency of CPSs and CPSoSs make this approach quite interesting. This is nowadays a

22 — Example #3: Neuroevolution

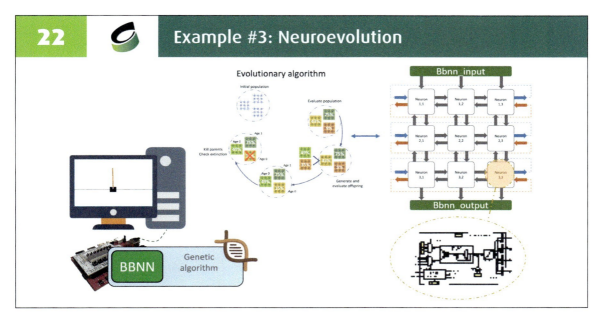

Finally, let see a third example, which combines artificial networks and evolvable HW. Neuroevolution refers to using evolutionary computation with neural networks as fabrics. At our Research Center we have developed a mesh-like fabric on an FPGA that consists on neurons that can be customized as conventional neuroins (different weights, bias and activation function), but also, in connectivity. Connectivity with neighbours is obtained by bidirectional paths, and the way to change both parameters and structure is achieved by partial reconfiguration of the device, guided by an evolutionary algorithm. It has been trained using OpenAI models, such as reverse pendulum or similar, with good results.

23 — CERBERO CPS Self-Adaptation

- **Cross-layer Approach** → **CPS** & **CPSoS** level
- All elements in the *adaptation loop* are included:
  - **Monitoring**
    - Context-awareness → Multiple sensors + Sensor Fusion
    - Self-awareness → HW and SW tasks common monitoring infrastructure → PAPI
  - **KPI extraction** → PHY and CYBER runtime models
  - **Adaptation management** → Dynamic task management → SPIDER
  - **Adaptation fabrics**
    - HW adaptation → mixed-grain, multiple solutions
    - ARTICo3, MDC, Just-In-Time composition, and mixed approaches
    - SW adaptation → Task migration between cores

23 CERBERO CPS Self-Adaptation

The next section presents some results achieved in the CERBERO Project. If is an EC funded project dealing ith CPS and CPSoSs adaptation. It addresses the different modules of the adaptation loop in a structured manner, allowing, for instance, SW, HW and heterogeneous fabric adaptation, using different reconfiguration engines, models and consistent monitoring schemes. Tools to help on the design of adaptable systems are a key result of the project. Academic partners have offered different instruments, such as ARTICo3, an architecture and toolset for dynamic management of slot-based HW accelerators, MDC (a corse-grain reconfiguration approach for HW adaptivity), dynamic schedulers for heterogeneous platforms such as Spider, or consistent monitoring via PAPI, an interface for both SW monitors which has been extended to HW and with automatic instrumentation via PAPIFY tool. The design flow is based on PREESM, a tool for describing tasks as actors such as in dataflow models of computation.

24 CERBERO multi-layer adaptation. CPS and CPSoSs loops

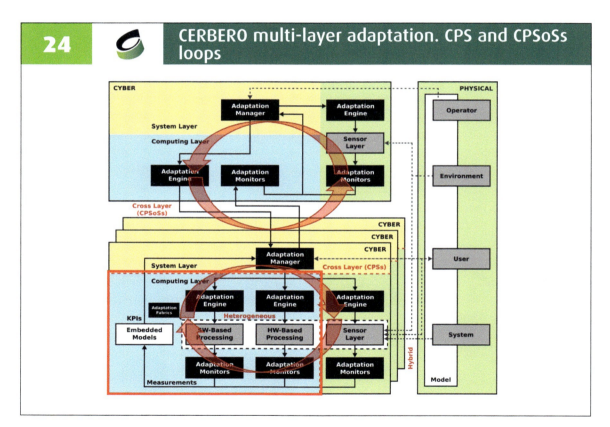

The adaptation loops have been identified at two levels: at individual CPS level and at CPSoS level. The image shows both loops in a hierarchical manner, identifying both layers, identifying heterogeneity (HW, SW and mixed adaptivity), and showing the dependencies with the environment (the physical part of CPS), showing he triggers for adaptation (self-monitors, sensors for external monitoring, and external triggers produced by the upper layer).

In the next slides we will talk about specific solutions for reconfigurable fabrics, monitoring of mixed HW/SW systems, execution modelling and an overall picture of results achieved in CERBERO (not all shown here).

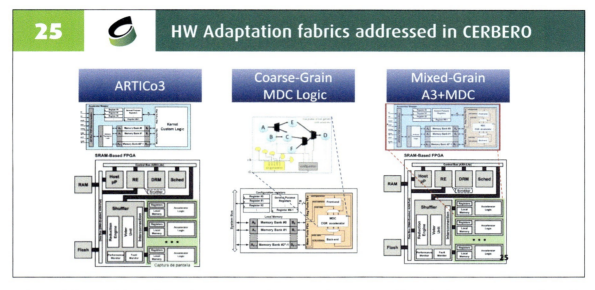

Three different fabrics have been targeted in CERBERO: ARTICo3 , MDC and a combination of both, for mixed-grain configuration. ARTICo3 is an architecture, a design tool and a run-time support library for HW acceleration. Multiple accelerators may be set into the programmable logic of the FPGA. The accelerators for a given task may operate in parallel mode or in redundant modes (dual module redundancy, DMR, or triple module redundancy, TMR). The design tools may start with VHDL descriptions of the accelerator kernel or via HLS synthesis. The accelerator kernel is wrapped with access logic to make it compatible with the ARTICo3 reconfigurable regions in the FPGA.

MDC (Multi dataflow composer) is a tool to generate coarse-grain architectures. Blocks are changed using virtual circuit reconfiguration, so that common functions are kept once in the logic and different functions can be selectively connected to the rest.

The mixed approach embeds MDC circuits within ARTICo3 slots, gathering the advantages of both adaptation/reconfiguration approaches.

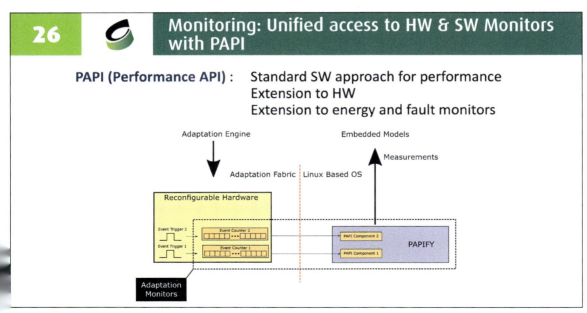

26 Monitoring: Unified access to HW &SW Monitors with PAPI

The monitoring aspect has been achieved by providing a consistent access method to monitors. Using as a starting point the PAPI (Performance API) library, used normally in SW, it has been extended with HW components that extend monitoring to all types of fabrics available in CERBERO. This way, SW performance monitors, HW performance monitors, energy/power monitors and fault counters for accelerator slots when operating in redundant execution modes are accessible with the same type of interfaces. A visualization tool, PAPIFY_VIEWER has also been developed to graphically monitor the fulfilment of real-time constraints, execution of tasks, scheduling, etc., as well as associated parameters such as energy consumed per task.

27 Modelling: Energy vs Execution time vs fault Tolerance

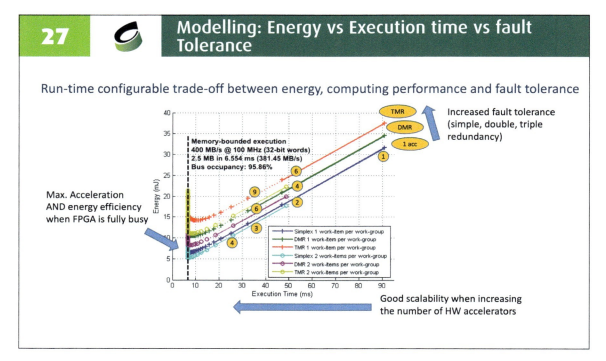

This graph shows a model to define the design space of ARTICo3 HW accelerators. The three lines represent the execution time and resulting energy consumption when operating in single, double or triple redundancy modes. The numbers in circles represent the points where multiple accelerators are working in parallel to further accelerate. It can be seen that, for the example at hand, good scalability in terms of accelerators is achieved until the bus that exchanges data to/from the accelerators gets saturated. As it can be seen, the optimum point is to have all accelerators busy, before the bus gets saturated. The trade-off between fault tolerance and acceleration is clear. For instance, if 1 accelerator is used, execution time is around 90 ms and energy around 32 mJ. If three accelerators were used instead, time could be reduced, sround 33 ms and 13 mJ, with no increased fault tolerance, or, if desired, TMR could be set, with additional energy consumption (37 mJ), no acceleration wrt 1 accelerator, butr complete fault masking.

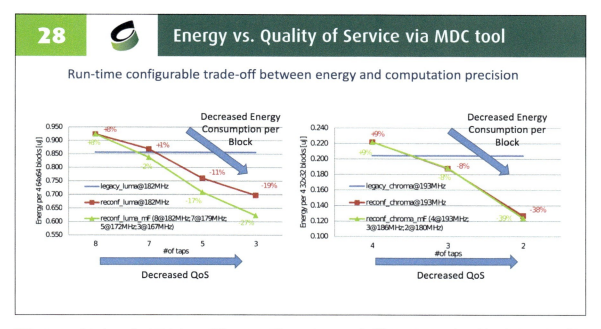

These graphs show, for MDC, how different quality results for an adaptive filter may be obtained by the use of different taps selectable via MDC, trading them off against different energy utilization.

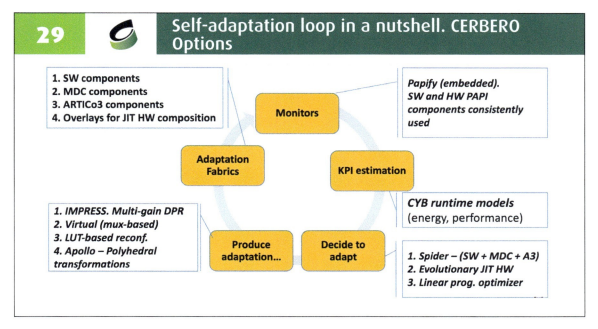

This slide presents, as a summary, all adaptation modules for CPS level developed in CERBERO. Apart of the before mentioned techniques, architectures, models and tools, work has been done in terms of multiple grain dynamic and partial reconfiguration, HW overlays for just-in-time HW composition, tools for SW modification based on Apollo polyhedral transformation tool, optimizers for linear programming, heterogeneous execution dynamic schedulers, ... The project aims at gathering the benefits of a structured adaptation-loop based approach.

HETEROGENEOUS CYBER PHYSICAL SYSTEMS OF SYSTEMS

30 Towards more reliable CPSs

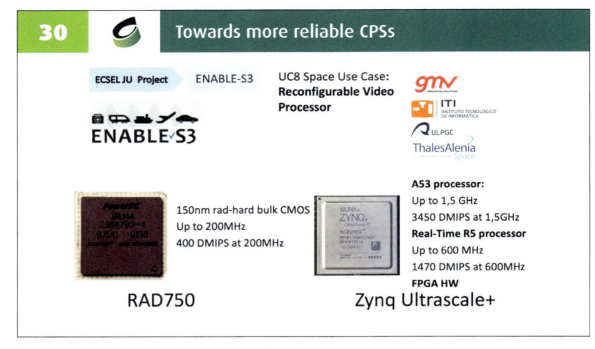

In this last section, the use of DPR for more reliable systems operation are presented. These are the results of ENABLE-S3 project, where, at CEI, in conjunction with other partners from the space sector, and funded by ECSEL JU and National Spanish R&D Plans, the possibilities of using SRAM-based FPGAs with increased reliability via design methods and fault mitigation techniques, have been explored.

Today's systems in space require more performant solutions than those achievable with radiation tolerant ones, and at a much smaller cost. The aim is to increase reliability by the architecture design, accounting with a combination of fault mitigation techniques. In the project, a complete verification and validation platform, based on a Zynq UltrScale+ FPGA was developed.

31 Reconfigurable Video Processor

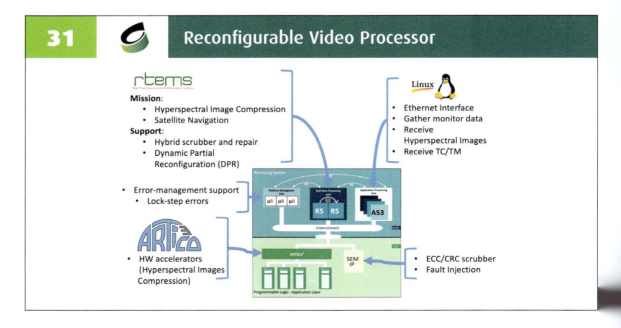

31 Reconfigurable Video Processor

The picture shows all the elements included in the Reconfigurable Video Processor Board developed. The Zynq UltraScale+ includes several processing fabrics. The R5 processors are run in lock-step mode, a fault-tolerant mode with spatial and temporal redundancy. Therefore, this dual-core executes as a single-core. On it, the RTEMS real-time operating system was ported, allowing for precise timing of tasks. There are two types of tasks: mission related ones and support related ones. Support for reconfiguration, fault detection and mitigation/repair is achieved by a combination of scrubbers that have been set to be compatible with reconfigurable HW. Fault isolation is combined with the capability of ARTICo3 to work in TMR (shown before). The slots have been used to accelerate mission tasks, such as hyperspectral image compression or satellite navigation.

The A53 cores are devoted for the test subsystem, and are intended to provide interfaces with other tools, such as activity monitors, interfaces with sensors that have been simplified (e.g., camera interfaces), external telecommand or telemetry, etc. Also, the Soft-Error Mitigation IP from the FPGA manufacturer peovides additional fault injection by ECC and CRC checking and repair, as well as an interface for fault injection.

32 Validating the Hardening of MPSoCs for Space Applications

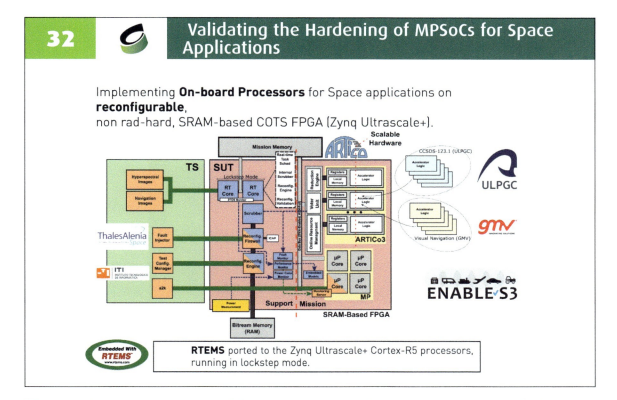

The overall system components of the system under test (SUT) and test subsystem (TS) are represented together in this picture. The complexity of the combined TS/SUT system is compensated by very fast RTEMS activity monitoring (using shared memory exchange regions between RTEMS for the SUT and Linux for the TS. We are now in the process of using this infrastructure to test the reliability and availability (reliability taking into account the repair capability) for the platform with diverse algorithms.

This is an example on how adaptivity may also be used for enhance reliability in critical systems.

33 Conclusions

- The world is becoming complex, and systems might need to be autonomous and adapt to changes w/o external intervention
- Full autonomy is obtained as a contribution of other interesting characteristics
    - DPR, scalability, self-awareness, self-healing, …
- Performance goes together with energy efficiency and dependability
- Do not forget security, safety, dependability, and other -ilities
- FPGAs and DPR are key players, in CPS and HPC domains
- There are many things left to be done → research opportunities
- Always try to identify the adaptation loops. A structured way of designing adaptation support may help

To conclude, adaptation is a very beneficial feature to comply by CPSs and CPSoSs in order to have a longer operation life. The ultimate levels of adaptation are self-adaptive systems, which are a must for fully autonomous unattended operation. Self-* properties are desirable in many systems, and they all have a common background based on the adaptation loop (knowing, measuring, deciding and performing adaptation on a fabric).

The use of adaptive systems not only opens up the possibilities for functional adaptation, but for extra-functional parameter adaptation as well. Several examples shown along this lecture have been used to show the trade-off between performance, energy consumption and dependability.

Adaptation requires additional features, not shown here, such as how to produce adaptation in a safe and secure manner. This is important for both worlds: CPS systems level (including IoT and edge processing) as well as for the High-Performance world.

This is an open field and there are many chances and research opportunities. From here, the research community is encouraged to continue the work in this field, and companies are encouraged to incorporate adaptivity and various self-* features in their products as an asset that ensures longer life-cycle in their products.

34 Acknowledgements

- Many things to be shown here today would not have been possible without the help of …
    - Teresa Riesgo, Jorge Portilla, Andrés Otero, Alfonso Rodriguez and Yago Torroja
    - Yana Krasteva, Wei He, Rubén Salvador, Javier Mora, Juan Valverde, Filip Veljkovic, Arturo Pérez, Leonardo Suriano, Borja Revuelta, Rafa Zamacola
    - Ángel Gallego, Blanca López, Julio Camarero, Santiago Muñoz, Alberto Ortiz, Alberto García
    - Miguel de las Heras, Ángel Morales, Manu Llinás, Alex Fernández, Miguel López, Pablo Sánchez, Carlos Pizarro, Verónica Díaz, Clara Casas, Miguel Baquero, Antonio Niño, César Castañares, Javi Vázquez, Javier Polop, Diego Lanza, Carlos Giménez, Victor López, Ramón Conejo, Carlos Correa, Pablo Iglesias, David Gozalo
    - … thank you all for being a great (if not the best) team
- We also acknowledge the support from institutions
    - European Comission, Ministerio de Ec. y competitividad, Min. de Industria, Comunidad de Madrid
- Many thanks for the collaboration with past, present and future partners, and all those who shared the same illusions

CHAPTER 03

Blockchain in Supply Chain Management

Dr. Harry Manifavas
Institute of Computer Science
Foundation for Research and Technology –
Hellas

Dr. Ioannis Karamitsos
Rochester Institute of Technology
Dubai Campus, UAE

#		Page	#		Page
1.	Contents	64	31.	SECMARK Interactive Blockchain Simulation (III)	80
2.	Blockchain in Supply Chain Management	64	32.	SECMARK: Create Asset	80
3.	Goals of Supply Chain Management	65	33.	SECMARK: Asset Attribute	81
4.	SCM vs Logistics	65	34.	SECMARK: Transfer the Asset (I)	81
5.	SCM: Key Challenges	66	35.	SECMARK: Transfer the Asset (II)	82
6.	The Role of Ledgers in Business	66	36.	SECMARK: Transfer the Asset (III)	82
7.	Blockchain as a Distributed Ledger	67	37.	SECMARK: A Blockchain Transaction (I)	83
8.	Tokenization	67	38.	SECMARK: A Blockchain Transaction (II)	83
9.	SC: Challenges Revisited	68	39.	SECMARK: A Blockchain Transaction (III)	84
10.	Blockchain Supply Chain (I)	68	40.	SECMARK: Selling a Component (I)	84
11.	Blockchain Supply Chain (II)	69	41.	SECMARK: Selling a Component (II)	85
12.	Blockchain Oracles	69	42.	SECMARK: Selling a Component (III)	85
13.	Blockchain Oracles: Three Types	70	43.	SECMARK: Transaction Recorded	86
14.	Blockchain Oracles - Challenges	70	44.	SECMARK: Add Information (I)	86
15.	Blockchain in SCM Benefits	71	45.	SECMARK: Add Information (II)	87
16.	Case Studies	71	46.	SECMARK: Add Information (III)	87
17.	Cargill: Traceable Turkey	72	47.	SECMARK: Integrator (I)	88
18.	Carico Café: Traceable Coffee	72	48.	SECMARK: Integrator (II)	88
19.	Everledger: Tracking Diamonds, Mine to Consumer	73	49.	SECMARK: Integrator (III)	89
20.	Walmart: Traceable Food Supply Chain	74	50.	SECMARK: Integrator (IV)	89
21.	DNV GL MyStory: Track Provenance of Products	74	51.	SECMARK: Assignation (I)	90
22.	Blockchain Supply Chain Management Demo	75	52.	SECMARK: Assignation (II)	90
23.	Supply Chain Management Blockchain Demo	75	53.	SECMARK: Assignation (III)	91
24.	SCM Blockchain Demo: Platform & Entities	76	54.	SECMARK: Decommissioning (I)	91
25.	SECMARK Interactive Blockchain Simulation (I)	76	55.	SECMARK: Decommissioning (II)	92
26.	SECMARK Interactive Blockchain Simulation (II)	77	56.	SECMARK: Decommissioning (III)	92
27.	Blockchain Demostration Site: Dashboard	78	57.	SECMARK: Disposal (I)	93
28.	Blockchain Demostration Site: Blocks	78	58.	SECMARK: Disposal (II)	93
29.	Blockchain Demostration Site: Transactions History	79	59.	SECMARK: Disposal (III)	94
30.	Blockchain Demostration Site: AMA	79			

A supply chain is the connected network of all the entities involved in the manufacture and final distribution of a product or service. It starts with the delivery of raw materials from a supplier to a manufacturer and ends with the delivery of the finished product or service to the customer. Supply chain management is the management of the flow of goods and services in order to cut excess costs and better meet customer expectations. Although supply chains have existed for a long time, the way they are managed is not as efficient as expected. The use of blockchains in supply chain management offers a number of advantages. A blockchain is a constantly growing ledger which keeps a permanent record of all the transactions that have taken place, in a secure, chronological and immutable way and can be shared by all the entities in the supply chain. The combination of supply chain management and blockchain adds trust in partners' interactions, reduces paperwork, increases security as well as transparency of records and transactions. Such an enhancement, indeed cuts excess costs and better meets customer expectations.

1 Contents

- Why use Blockchain in Supply Chain management (SCM)?
- How Blockchain is used in SCM
- Benefits and Limitations of Blockchain
- Case Studies: Blockchain Use Cases in the Supply Chain
- SCM Blockchain demo

Supply chain management brings all stakeholders (from raw materials to products) together. Each party has a special role in the supply chain. The stages involved are:
- Raw Material
- Supplier
- Manufacturing
- Distribution
- Customer

With blockchain we achieve efficiency and speed in operations as well as tracking of all goods.

2 Blockchain in Supply Chain Management

- *We estimate that the application of blockchain to global supply chains alone could result in more than $100 billion in efficiencies*
- *Add improvements in provenance and traceability of pharmaceuticals and food*
  - Ginni Rometty
  - Chairman, President, and CEO, IBM
  - https://www.ibm.com/ibm/ginni/01_09_2017.html
- Main benefits
  - Transparency
  - Efficiency
  - Security

Blockchain is a disruptive technology with potential applications in the sustainable sharing economy. One of these areas is Supply Chain Management. It is estimated that the applications of blockchain to global supply chains could result in more than $100 billion in efficiencies.

With the usage of blockchain we experience improvements in provenance and traceability in many areas, like food and pharmaceuticals, for example. Provenance is the ability to track any food or pharmaceutics used for consumption, through all the stages, from production to distribution. Traceability is a way of responding to potential risks that can arise to ensure that all food and pharma products are safe for people.

Main benefits of using blockchain technology are transparency, efficiency and security between the parties and associated transactions.

3 Goals of Supply Chain Management

In a supply chain a number of entities need to collaborate.

Suppliers and vendors provide raw materials to manufactures.

Manufacturers transform raw materials into product to sell.

Finally, when the product is ready, the distribution channels move the product into the market in a number of ways (wholesale, retail, online).

- **Fulfill demand efficiently**
  - "You can't sell from an empty wagon"
- **Drive customer value**
  - Products get to the right customer at the right time
  - Happy customers are repeat customers, they keep growing your business
- **Improve responsiveness**
  - Despite hiccups affecting normal operation
- **Contribute to financial success**
  - SCM integral part of business

The main goals of Supply Chain Management are to fulfill the demand in an efficient way, drive the customer value in order to reduce the churn and contribute to financial success for every member in the chain.

4 SCM vs Logistics

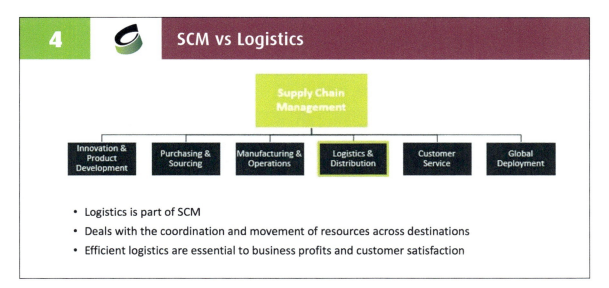

- Logistics is part of SCM
- Deals with the coordination and movement of resources across destinations
- Efficient logistics are essential to business profits and customer satisfaction

The term SCM and Logistics many times are confused and used interchangeably. SCM is a concept that links together multiple processes to achieve a competitive advantage, while logistics refers to the movement, storage, and flow of goods, services and information. As we can see the SCM incorporates a set of subprocesses such as logistics and others.

The most important difference is that in logistics a single organization is involved and customer satisfaction is the target while in SCM multiple organizations are involved and competitive advantage is the target.

5

5 SCM: Key Challenges

- **Multiple separate players**
  - May be around the world
- **Lack of transparency**
  - Each contains and controls a certain amount of data
- **Discrepancies in records**
  - Subject to tampering
- **Limited cross process visibility**
  - It's important to have the big picture
- **Globalization**
  - Subjected to events at other parts of the world
- **Maintaining high performance**
- **Counterfeiting**

During the implementation of SCM, the key challenges present are: multiple separate players, lack of transparency, discrepancies in records, limited visibility, globalization, high performance and counterfeiting. With the use of blockchain all these challenges will be met either by introducing improvements in the process or by eliminating the causes altogether.

With blockchain counterfeiting is eliminated, due to the fact that any transaction or document is immutable, meaning that one cannot alter or delete any part of the information/asset. Counterfeiting applies to the healthcare sector as well. According to WHO, 10-30% of all medicine in developing countries is counterfeit. With blockchain's traceability property one is able to figure out the origin of manufacturing and all the parties involved in the supply chain.

6 The Role of Ledgers in Business

- Each market defines its ledgers
- Ledgers record business activity and track transfer of assets
- Ledgers are everywhere
- Each party keeps separate ledgers
- Discrepancies in ledgers
- Errors, inefficiency, manipulation

The idea of a Ledger has been around since the ancient times. Every business transaction and tracking of assets were recorded in the ledger of each participating entity. Each entity creates and keeps its own ledger. A ledger is like book that records every transaction or transfer of assets is its pages. Its clear that due to human intervention many ledgers face discrepancies and input errors.

7 Blockchain as a Distributed Ledger

- One ledger shared by all parties
- Provides a single source of truth that is verifiable, tamper-proof and unchangeable (digital signatures)
- Creates a record of any transaction of value, whether it be money, goods, property, work or votes
- Replicated, distributed and validated by consensus
- Has the potential to transform how humans transact

Blockchain is based on the distributed ledger concept. Each participant in the blockchain network has access to a shared ledger which acts as a recording mechanism for transactions such as exchange of assets and data among network participants. It is a database that it shared and is synchronized among the members of a decentralized network. Any transaction is validated by consensus protocols and all copies of the ledger contains the same information.

8 Tokenization

- The process of representing an asset as a token that can be stored, recorded and transferred on a blockchain
- Transferring a physical asset
  - A unique token per asset
  - Includes metadata for the item
    - Provenance
    - Certifications
    - History
    - ...
  - Unique identifying QR code

A digital token can be any kind of digital asset or any digital representation of a physical asset. Also, a digital token can represent any fungible, tradable good, such as currency, reward points or gift certificates. With tokens we can assign a value to each asset that can be stored, recorded, and transferred on a blockchain. Each asset is associated with a unique token which includes metadata for the asset such as provenance, certificates and history.

9 SC: Challenges Revisited

- **Separate "legacy" systems**
  - Supply Chains have been around for a long time → old systems in place
- **Silos of information**
  - Entities don't share information → no single point of reference
- **Paper based**
  - A farmer delivering avocados most probably handles everything on paper
- **Redundancy and duplication of efforts**
  - Filling in multiple forms
- **Discrepancies**
  - Recording of information may not be consistent
- **Vulnerable to tampering**
  - Some people may record incorrect information

During the SC chain implementation, we found many challenges and issues that we need to taken into consideration.

Many organizations use legacy IT systems, tools and software negatively affecting the time taken for the execution of orders. Organizations don't share information with other participants and that makes tracking of assets harder. It's not uyncommon for the whole process to be based on paper. Paper forms to be filled are too many and there is always the potential for human errors. As a consequence, error correction takes time and is costly.

10 Blockchain Supply Chain (I)

- **Shared permissioned blockchain**
  - Who gets access to what information
  - However, the business has full access to all information → can see the big picture
  - Cryptographic primitives are used for signatures, authentication, etc.

- **Can track individual assets on the chain (e.g., by assigning unique QR codes)**
  - Can provide the customer with an option
  - A unique QR code will allow the customer to see how the product was made and its way through the supply chain

The aforementioned challenges can be met by the use of blockchain. All stakeholders can become blockchain nodes sharing the same ledger. Each ledger contains all the details in electronic forms and tracking is easy for all the parties. Information is crosschecked and then saved. Afterwards it cannot be modified. This approach eliminates errors.

One way to track individual assets on the chain is by using QR codes. These codes contain all the information required in order to track an asset.

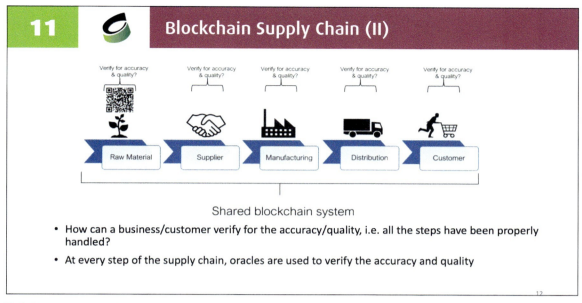

When a blockchain is used, each member of the chain has all the information that is permitted to know. One can then ask each party different questions for the asset. We introduce the concept of oracles for the verification, accuracy and quality of the asset.

Blockchain Oracles

- An oracle is a data feed designed for use in smart contracts on the blockchain
  - Oracles provide external data and trigger smart contract executions
  - Smart contracts contain value and only unlock that value if certain pre-defined conditions are met
  - Such condition could be any data like weather temperature, successful payment, price fluctuations, etc.
  - When a particular value is reached, the smart contract changes its state and executes the programmatically predefined algorithms, automatically triggering an event on the blockchain

Blockchains cannot access data outside their network, so they need a 3rd party provided service to be able to do so.

An oracle, in the context of blockchains and smart contracts, is an agent that finds, verifies and submits real-world information to a blockchain to be used by smart contracts. A smart contract is triggered by an external agent-oracle providing all the external data to the blockchain network.

Oracles come in two forms. They can be "callable" from the smart contract by generating an event with a request/reply interaction or they can be "trigger"-based where the oracle calls the smart contract when certain conditions are met. To include a third-party oracle in the smart contract, it needs to be trustworthy and reliable since its interaction with the blockchain is immutable and hence irreversible.

HETEROGENEOUS CYBER PHYSICAL SYSTEMS OF SYSTEMS 70

13 Blockchain Oracles: Three Types

- Hardware
  - Triggered by a sensor (RFID/IoT sensor, barcode scanner, etc.)
  - Tag fish to track its way through the supply chain

Image Credit: WWF   Image Credit: Provenance   Image Credit: FEMA Photo Library   Image Credit: Fraunhofer-IPMS

- Software
  - Information available online (e.g., market data, data triggers, etc.)

- Human (inspectors)
  - Verify and research events (e.g., quality control, special knowledge, etc.)

Hardware oracles: This form of oracle is sending data to a smart contract as a result of an event in the physical world. For example, in supply chain management, if an asset/object with an RFID tag was to arrive at a particular warehouse, this data can then be sent to a smart contract. As a result, hardware oracles can facilitate the tracking of goods along the supply chain.

Software oracles: This form of oracle typically includes online sources of information that are easily accessible, for example, market data, websites and public databases. Software oracles are the most powerful type of oracles because of their inherent interconnectedness to the internet; this connection allow software oracles to supply the most up-to-date information to smart contracts.

Human oracles: This form of oracle can verify and research events and is based on human interaction.

14 Blockchain Oracles - Challenges

- The primary task of oracles is to provide values to the smart contract in a secure and trusted manner
  - Third parties need to be trusted, they are not part of the blockchain consensus mechanism
  - It is critical that the source of the information is trustworthy (whether a website or a sensor)
  - Incorrectly recorded information cannot be changed
  - Providing smart contracts with trusted information sources is crucial for the users because in case of mistakes there are no rollbacks

- Sometimes, more than one oracle may be needed to trigger an action
  - Using only one source of information may be unreliable (outages, tampering, etc.)

- Different trusted computing techniques can be used as a way of solving these issues

15

Oracles are part of the blockchain and they need to provide data to the smart contract in a secure and trusted manner. To achieve this, all the external parties need to be trusted and the source of the information shall be trustworthy.

It is important for smart contracts to act on trusted information because in case of errors there are no rollbacks.

In some cases, more than one oracles are required to provide data to a smart contract.

15 Blockchain in SCM Benefits

- Immutable system of proof protected by cryptography and timestamps
  - Know exactly when each step is achieved

- Immutable nature incentivizes suppliers to input more accurate data
  - Input is signed so that others know who inputted incorrect information

- Increases level of trust in partners

- Reduced paperwork and redundancy of data input
  - Everybody is working with the same computer system

- Increased transparency across entire supply chain

Like any new promising disruptive technology the list of its benefits is long. With blockchain technology there is more security, immutability, and transparency of records and transactions. The combination of SCM and blockchain adds trust in partners' interactions, reduces paperwork and provides redundancy of data inputs. Moreover, the data stored in blockchain is protected from security threats using cryptography and timestamps methods.

16 Case Studies

- Blockchain in Supply Chain Management
  - Cargill: Traceable Turkey
    - https://www.youtube.com/watch?v=5dccy2s5uHE&feature=youtu.be
    - https://www.youtube.com/watch?v=E3SSIw7gt0w&feature=youtu.be
  - Carico Café: Traceable coffee
    - https://bitcoinexchangeguide.com/ugandas-carico-cafe-connoisseur-uses-blockchain-to-trace-coffee-shipments-and-production/
  - Everledger: Tracking diamonds from mine to consumer
    - https://diamonds.everledger.io/
  - Walmart: Traceable food supply chain
    - https://bitsonline.com/walmart-food-suppliers-blockchain/
  - MyStory: Track provenance of wine and products
    - https://mystory.dnvgl.com/
    - https://brandcentral.dnvgl.com/mars/embed?o=3298BC5ABCED2098&c=10651&a=N

In the following slides we present a number of case studies where the blockchain technology has been applied. The first case is traceable turkey from the Cargill Organization. The second case is traceable coffee for Carico brand. The third case is diamond tracking from Everledger. The fourth case is food tracking from Walmart. The fifth case is wine tracking from Mystory.

17 Cargill: Traceable Turkey

- Farm-to-table blockchain traceability of turkey

- "Meet your farmer" is offered during Thanksgiving season
- Launched pilot in 2017 in Texas for 60000 Honeysuckle White turkeys
- Expanded to 200000 turkeys in 2018 from 70 farms, 1/3 of turkeys sold for that brand
- 3500 retail stores across USA during the 2018 Thanksgiving season

- Through a text message, or by entering an on-package code at honeysucklewhite.com consumers can trace their turkey back to the family farm
- Includes farm location, family farm history, photos from the farm and message from the farmer

Cargill is expanding its blockchain-based turkey traceability program to cover metro areas in 30 US states after receiving "overwhelming interest" from farmers.

The program launched in a small pilot offering consumers only in Texas more information about the farms 60,000 Honeysuckle White-branded turkeys came from, via text or the Honeysuckle White website.

18 Carico Café: Traceable Coffee

- Uganda based: Carico Café Connoisseur

- Blockchain traceability of coffee across the supply chain
- Farmers can digitally integrate an immutable certification, including a QR code
- First shipment of multiple tons of "Bugisu Blue" coffee in December 2018

- Blockchain traceability allows coffee farmers to charge more
- Customers are willing to pay more if they know exactly where it comes from

- Once coffee makes it to the final destination, consumers can verify authenticity
- Consumers also access data of when and where it was grown and its grade

18 Carico Café: Traceable Coffee

Carico Café, the Ugandan organization behind this innovative idea, claim that the technology will present the industry with a range of notable benefits. For example, it is claimed that those purchasing coffee beans are prepared to pay higher rates if they have the option to trace the coffee back to its original location. This will subsequently allow local coffee suppliers to charge more and as such, pay their farmer's higher wages. Due to nature of limited processing capacities in Uganda, the vast majority of its coffee exports must be in the form of raw beans. As such, it makes it difficult for the end-buyer to ascertain where the beans originated from.

Through the use of blockchain technology, Ugandan farmers will have the option to digitally integrate an immutable certification. Once the coffee makes its way to its final destination, consumers will be able to verify the authenticity of the certification by scanning a QR code with their smartphone device.

19 Everledger: Tracking Diamonds, Mine to Consumer

- Founded in 2015

- Tracks the provenance of high-value assets on a global digital ledger and record it on the blockchain
- Encrypted the provenance of over two million diamonds

- Work across diamond supply chain including manufacturers and retailers

- In 2017, developed the Diamond Time-Lapse initiative
- Historical ledger of the movement of diamonds with real-time data
- Includes diamond origin, cutting, polishing, master artisans' work & certification

Everledger's cutting-edge blockchain-based solution has been integrated with Brilliant Earth's supply chain in collaboration with Dharmanandan Diamonds (DDPL), to seamlessly and securely track diamond origin and provide greater consumer assurance and insight around responsible practices for a collection of diamonds. Customers can now purchase the diamonds which highlight if they are blockchain-enabled via a badge on the Brilliant Earth website. Customers can also filter specifically for blockchain-enabled diamonds or browse through a broad selection of blockchain- enabled diamonds to find their perfect Beyond Conflict Free DiamondTM.

Appealing to customers that are interested in learning more about how and where their products are sourced, Brilliant Earth's website now highlights the journey of blockchain-enabled diamonds, including additional information throughout the cutting and polishing process. For example, information on the diamond product page includes provenance, rough carat, weight, rough diamond videos, and planning images to aid the cutting process. Additionally, once a customer has purchased a blockchain-enabled diamond, all transfers of ownership are securely recorded in the blockchain.

20 Walmart: Traceable Food Supply Chain

- Blockchain tracking of food supply chain
    - Increased efficiency and improves supply chain transparency
    - Began tests in 2016

- Suppliers and products tracked on blockchain
    - All Walmart leafy-greens suppliers mandated to adopt solution by Q3 2019
    - Plans to extend to other fresh fruit and vegetable suppliers over the next year
    - Tracking pork products across China

- Manage contamination cases
    - 48MM people in USA become ill from food born diseases
    - 1% reduction in food born disease in USA = $700B in savings
    - Reduced time to track produce from average 6 days to under 3 seconds

- Uses IBM Food Trust blockchain platform (Hyperledger Fabric)

Walmart is a retail food marketplace which has implemented a decentralized solution using blockchain for tracing the food products offered in its stores. This food traceability system is based on Hyperledger Fabric. The system tracks suppliers and products. A significant benefit of this system is that now Walmart can manage in a much more efficient way most contamination cases.

21 DNV GL MyStory: Track Provenance of Products

- DNV GL launches My Story™ - the blockchain based solution to tell the product's full story
    - https://www.dnvgl.com/news/dnv-gl-launches-my-story-the-blockchain-based-solution-to-tell-the-product-s-full-story-113549

- Blockchain tracking of wine and products
- Mobile app scan QR code

- "Grape-to-Bottle" visibility (full history)
- Quality, authenticity, origin, ingredients
- Water and energy consumption

- Currently, three Italian wine makers featuring labels
- Uses VeChain, an Ethereum-Based Solution

The Italian company DNV GL has launched MyStory, a blockchain-based solution for tracking wine. Mystory is a solution that allows consumers an easy glance into the contents of each wine bottle. Companies can now increase the transparency in their supply chain and demonstrate what lies behind the unique characteristics of their product. All this information is retrieved using a smartphone application to scan a QR code found on the wine bottle.

22 Blockchain Supply Chain Management Demo

Blockchain

Supply Chain Management

Demo

This application is designed to demonstrate how the life cycle of an asset can be modelled on a blockchain.

The secmarking blockchain demo platform is used with permission.

23 Supply Chain Management Blockchain Demo

- Securemarking has produced an interactive, supply chain focused blockchain demo

- Blockchain for Supply Chain Management Classroom Demonstration
    - http://blockchain.securemarking.com/

- User Guide
    - http://blockchain.securemarking.com/guide.html

- Instructional videos
    - https://www.youtube.com/watch?v=-yq0Z5Lif6Q&t=0s&list=PLZCE4KECNoHwpfb2rv49Tn7jtw06EsS1w&index=7

This demo platform is an educational project sponsored by the Airforce Institute of Technology with Additional help from the Beacon School of Business at University of South Dakota. It uses unclassified, fictitious data in a simple fictitious military scenario. It does not reflect official policy or position of US Air Force, Department of Defense or US Government.

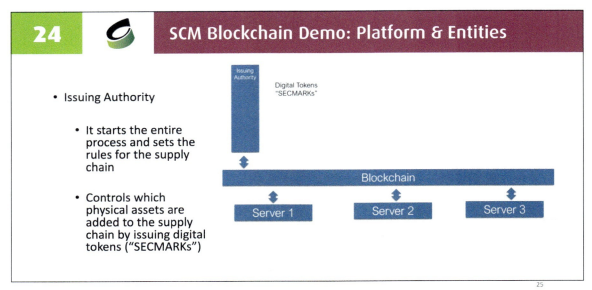

The demo platform is built on a private, fully centralized permissioned blockchain. It operates on three servers in a cloud computing environment. A number of different rules apply in the supply chain.

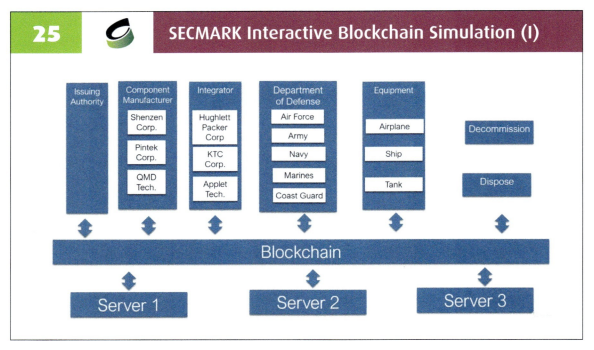

The three Component Manufacturers (Shenzen Corp., Pintek Corp., QMD Tech) need "SECMARKs" from the Issuing Authority before selling components.

The three Integrators (Hughlett Packer Corp., KTC Corp., Applet Tech) acquire components from the manufacturers.

DoD purchases products from the integrators for 5 branches (Air Force, Army, Navy, Marines, Coast Guard) and assigns products to 3 types of equipment (Airplane, Ship, Tank)

At the end, components need to be decommissioned and disposed of.

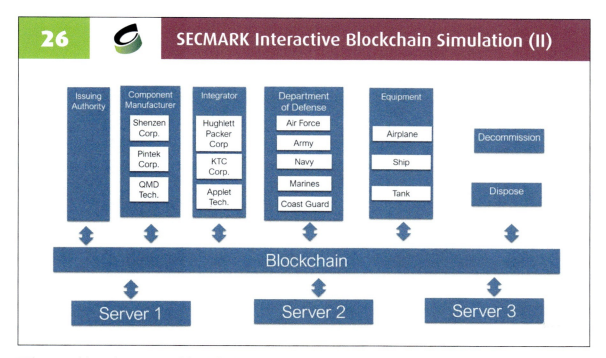

The asset life cycle consists of the following stages:
- The Asset Management Authority creates an asset template.
- The asset template is transferred to the Component Manufacturer.
- The Component Manufacturer updates the asset PIN, Product, Name, Assembly and SKUid attributes.
- The asset is transferred to the Integrator (OEM).
- The Integrator updates the asset Assembly attribute.
- The asset is transferred to the Department of Defense.
- The Department of Defense transfers the asset to some equipment.
- Upon completion of the assets useful life, the asset is transferred to the Decommissioning Authority.
- The Decommissioning Authority disposes the asset.

27 — Blockchain Demostration Site: Dashboard

On the real-time Dashboard you can see the blockchain in operation. Output shown in Transaction History depends on user access permissions.

28 — Blockchain Demostration Site: Blocks

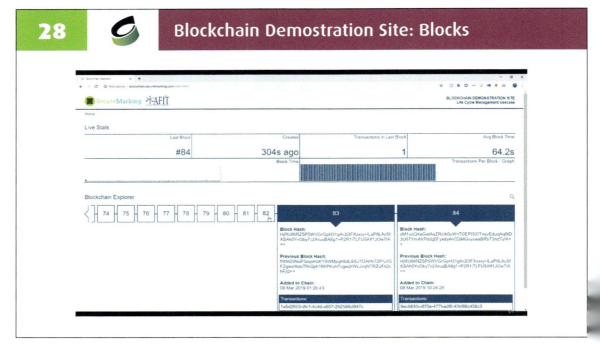

By clicking on a block you can see the details of that block (Block Hash, Previous Block Hash, Transactions).

29 Blockchain Demostration Site: Transactions History

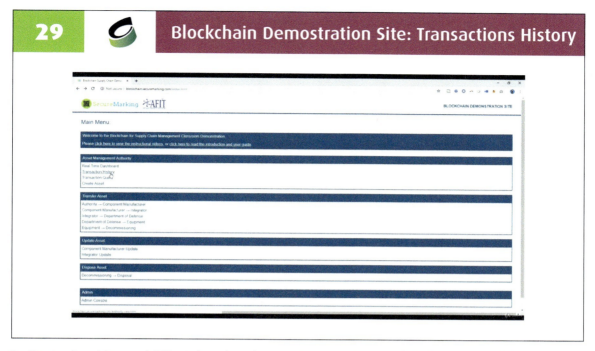

In Transactions history, visibility is based on the permissions assigned to individual parties. Different parties can view different transactions. For example, you can reduce it to a specific component manufacturer.

30 Blockchain Demostration Site: AMA

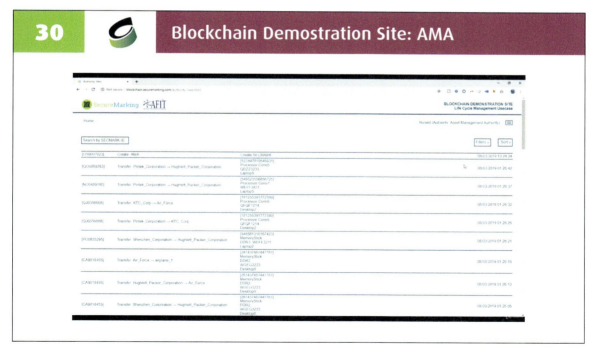

The Asset Management Authority can view all transactions. The rest of the participants can only see transactions of assets they've owned, up to the point where they transfer them to the next participant.

HETEROGENEOUS CYBER PHYSICAL SYSTEMS OF SYSTEMS

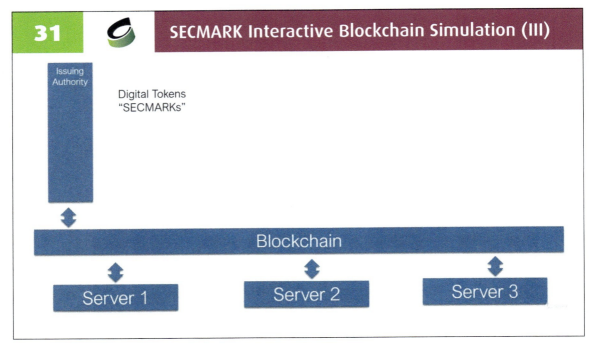

Step 1:
Asset and digital token creation

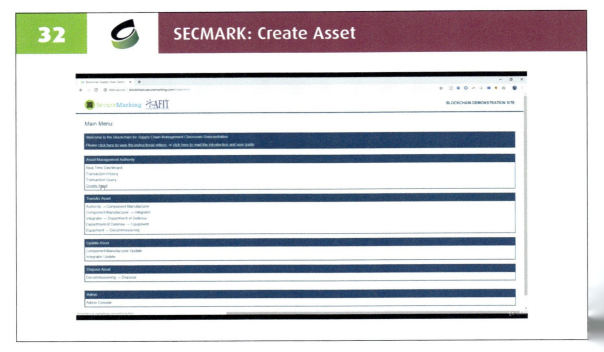

Step 1:
Click Create Asset to create SECMARK and the system will come up with a unique serial number.

33 SECMARK: Asset Attribute

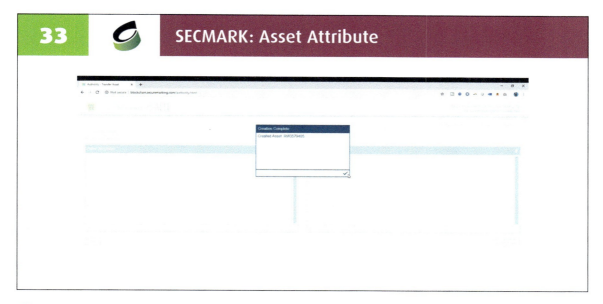

Step 1:
Each asset attribute is written to the blockchain. This means once an asset attribute is committed to the blockchain, it cannot be changed by any participant. This is due to the immutable nature of blockchains.

Each asset attribute has its own access permissions which are defined in the blockchain. The participants can read all attributes for assets they own, however they can only change the attributes they are permitted to through the blockchain.

34 SECMARK: Transfer the Asset (I)

Step 2:
Transfer the asset to Component Manufacturer (QMD Tech. in our case). The Component manufacturers need the SECMARKs from the issuing authority before they can sell components further down the stream.

35 SECMARK: Transfer the Asset (II)

Step 2:
Click on the + sign to select Asset and Component Manufacturer. Then, click Transfer Assets.

36 SECMARK: Transfer the Asset (III)

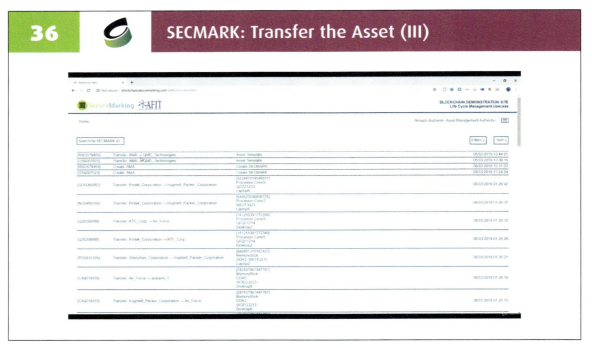

Step 2:
Blockchain recorded the transaction. This transfer can be verified in Transaction History.

37 — SECMARK: A Blockchain Transaction (I)

Step 3:
The Component Manufacturer may update / add information to the asset. This is a blockchain transaction.

38 — SECMARK: A Blockchain Transaction (II)

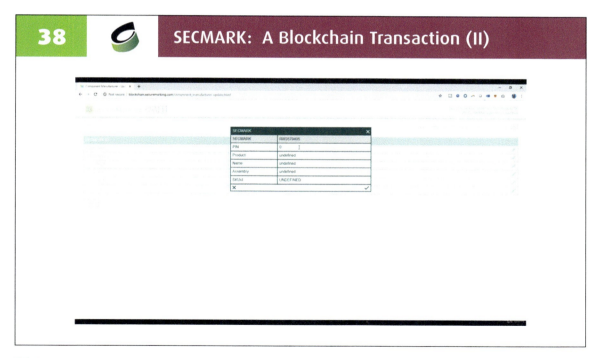

Step 3:
Changes are made to the following fields: PIN, Product, Name, Assembly and SKUid.

39 — SECMARK: A Blockchain Transaction (III)

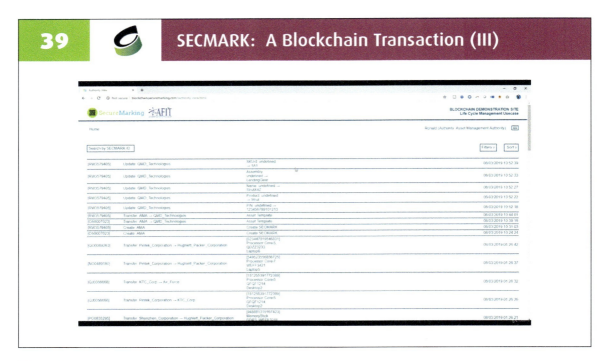

Step 3:
Updates are committed to the blockchain which recorded the transaction. This update can be verified in Transaction History.

40 — SECMARK: Selling a Component (I)

Step 4:
The Component Manufacturer sells Component to Integrator (Hughlett Packer Corp. in our case).

41 SECMARK: Selling a Component (II)

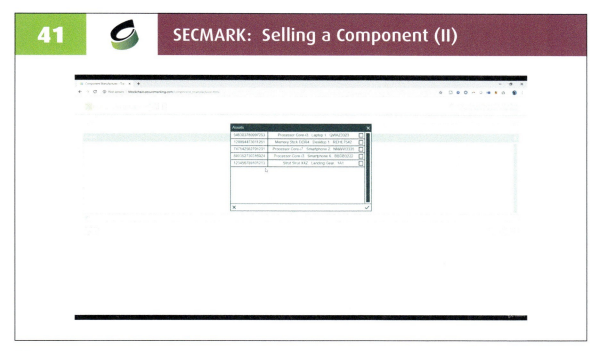

Step 4:
  Select asset.

42 SECMARK: Selling a Component (III)

Step 4:
  Select Integrator. Then, click Transfer Assets.

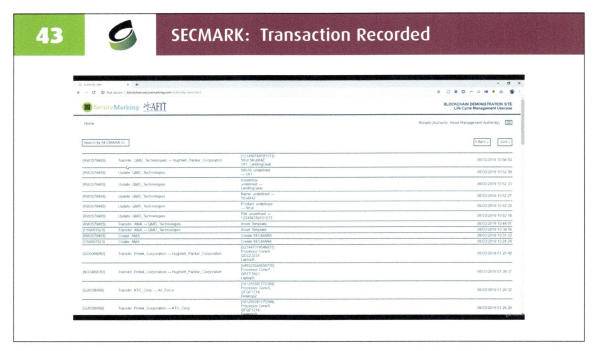

Blockchain recorded the transaction. This transfer can be verified in Transaction History.

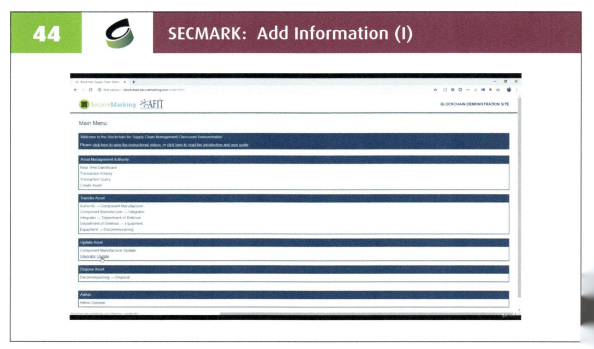

Step 5:
The Integrator can add information to the asset. This is also a blockchain transaction.

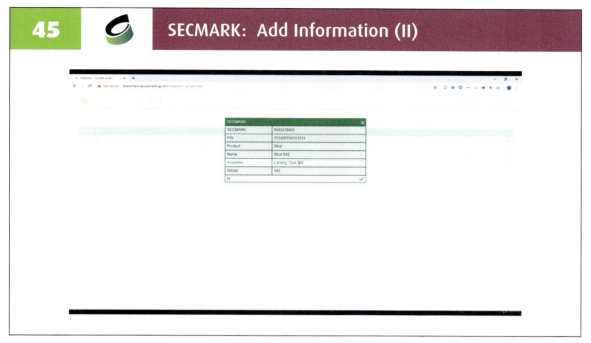

Step 5:
The Integrator can change some but not all information.

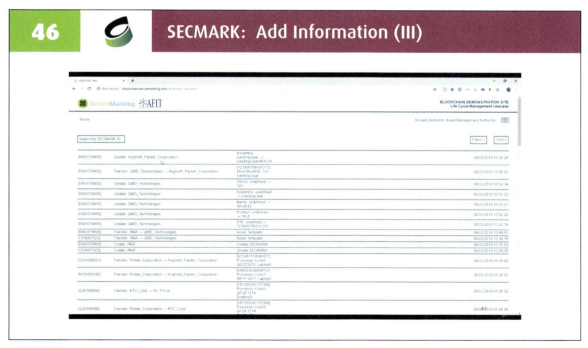

Step 5:
Blockchain recorded the transaction. This update can be verified in Transaction History.

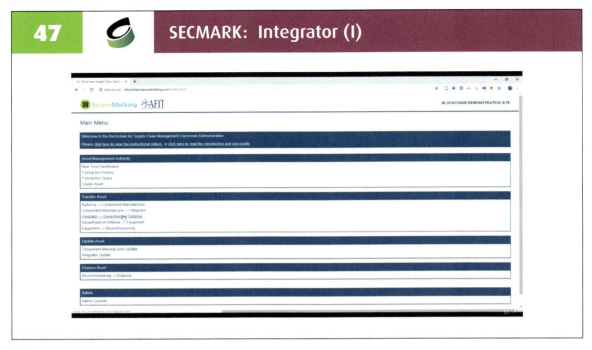

Step 6:
The Integrator sells its product to the DoD. This is also a blockchain transaction.

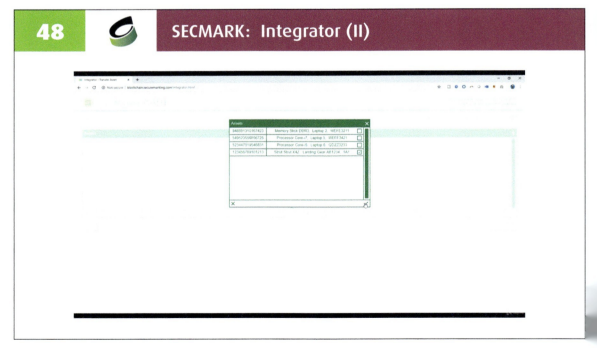

Step 6:
Select the right Integrator and then select the Asset.

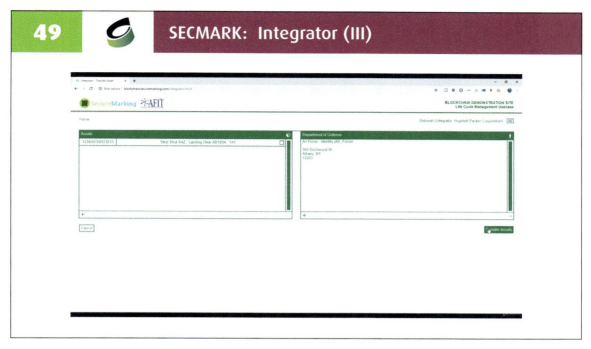

Step 6:
Select the DoD branch (Air Force in our case), Then, click Transfer Assets.

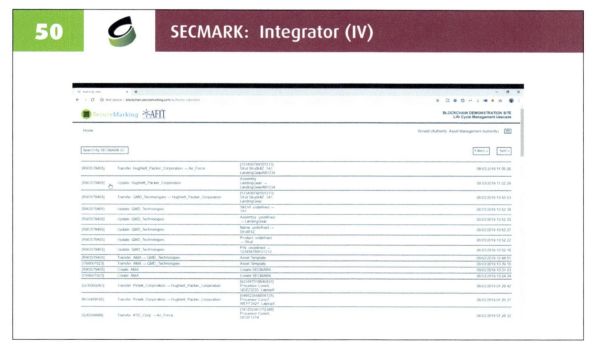

Step 6:
Blockchain recorded the transaction. This transfer can be verified in Transaction History.

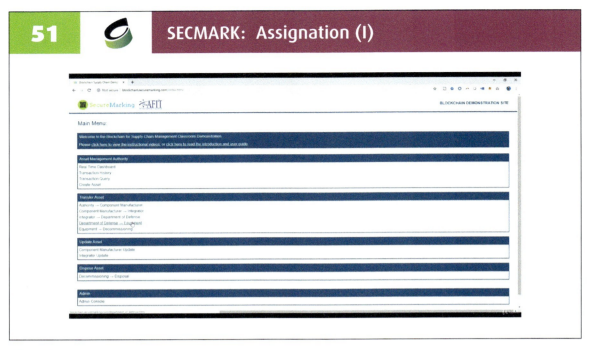

Step 7:
    DoD assigns this asset to a specific piece of equipment. This is also a blockchain transaction.

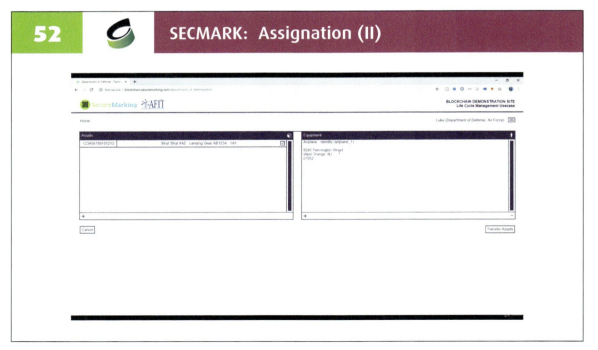

Step 7:
    Select Asset and Piece of Equipment (Airplane in our case). Then, click Transfer Assets.

53 — SECMARK: Assignation (III)

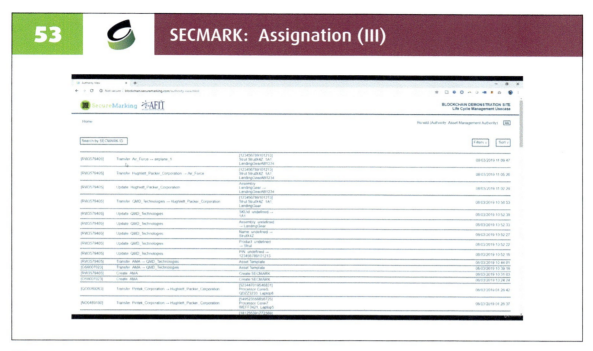

Step 7:
Blockchain recorded the transaction. This transfer can be verified in Transaction History.

54 — SECMARK: Decommissioning (I)

Step 8:
At end of its life, this piece of equipment needs to be decommissioned by the Air Force. Select equipment decommissioning. This is also a blockchain transaction.

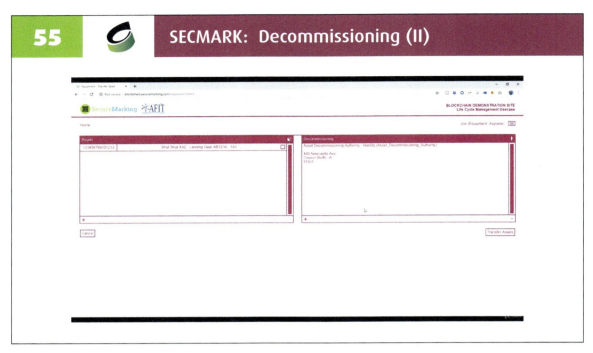

Step 8:
Select Asset and Asset Decommissioning Authority. Then, click Transfer Assets.

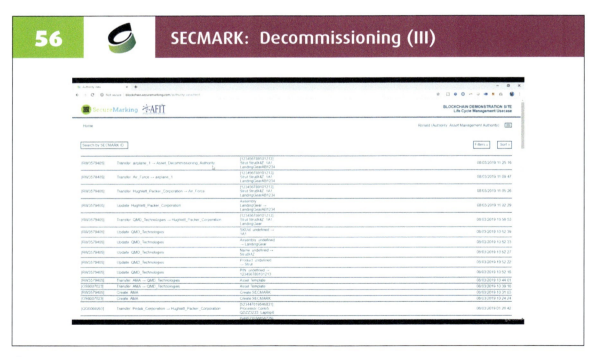

Step 8:
Blockchain recorded the transaction. This transfer can be verified in Transaction History.

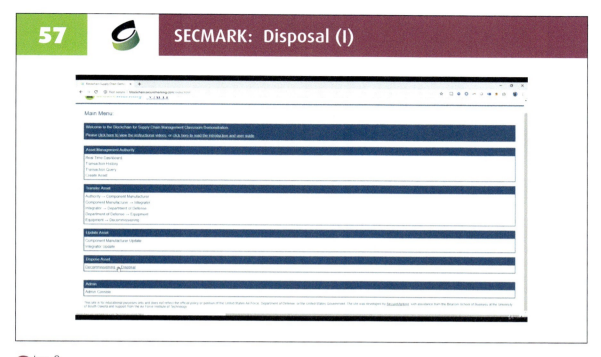

Step 9:
Eventually, the asset needs to be disposed of. Select Decommissioning à Disposal. This is also a blockchain transaction.

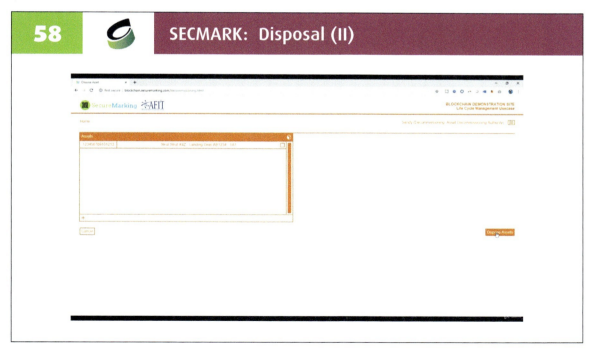

Step 9:
Select Asset. Then, click Dispose Assets.

59 SECMARK: Disposal (III)

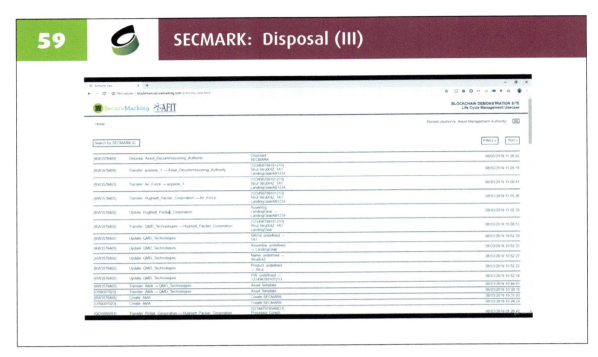

Step 9:
Blockchain recorded the transaction. This transfer can be verified in Transaction History.

CHAPTER 04

Edge Intelligence: Time for the Edge to Grow Up!

Prof. Theocharis Theocharides

University of Cyprus

#	Title	Page
1.	1964: Stanley Kubrick's "Connected World"	100
2.	KIOS Research and Innovation Center of Excellence - UCY	100

PART 1 — 101

#	Title	Page
3.	IoT: Complexity Trends and Challenges	101
4.	IoT: Complexity Trends and Challenges	102
5.	From the Cloud, to the Fog, towards the Edge…	102
6.	What does it mean in real-life IoT!	103
7.	Sensor Evolution	103
8.	Sensors Abundant in the IoT Era!	104
9.	Remote Sensing and Sensor Fusion	104
10.	What about the Actuators?	105
11.	So…	105
12.	But…the Senses & Muscles Need a Brain!	106
13.	Using Vision as an Example!	106
14.	Technology Scaling	107
15.	But we don't just go from this…	107
16.	Why Intelligent Edge Devices Become Critical	108
17.	Traditionally…	108
18.	Edge Intelligence	109
19.	Edge Intelligence (…continued)	109
20.	The Hype of Machine Learning	110
21.	Defining Artificial Intelligence	110
22.	What Makes it an AI SoC?	111
23.	AI's Insatiable Need for Compute, Memory, & Electricity	111
24.	Data Center AI is Moving to the Edge	112
25.	Architectures For Deep Neural Networks…	112
26.	The Race for the Edge is ON!	113
27.	Google's Edge TPU and NVIDIA Jetson TX2	113
28.	Why Edge Intelligence Matters…	114
29.	Deep Learning and Edge (IoT/SoC) Challenges – Nature vs. Nurture	114
30.	Specialized Processing Challenges in AI	115
31.	Deep Learning SoC Challenges	115
32.	Memory Bandwidth at the Edge	116
33.	Memory Performance Challenges in AI	116
34.	Deep Learning SoC Challenges	117
35.	Real-Time Connectivity Challenges in AI	117
36.	Securing our little Edges!	118
37.	Nurturing Amazing AI SoCs	118

PART 2 — 119

#	Title	Page
38.	Challenges and Opportunities of Edge Intelligence	119
39.	Challenges	120
40.	Complexity of Neural Networks	120
41.	Designing Neural Networks for Edge Intelligence	121
42.	Energy-Efficient Deep Learning	121
43.	Energy and Performance-Efficient DNNs	122
44.	Robust Deep Learning	122
45.	Brain-Inspired Computing: Trends	123
46.	50 Years of Algorithms Research – What about the Memory?	123
47.	Emerging Memory Technologies	124
48.	Emerging Memory Technologies	124
49.	Challenges Associated with ML @ the Edge!	125
50.	Computational Efficiency Gap: AI vs. Humans	125
51.	Why Brain-Inspired Computing…	126
52.	…for Big Data? Inspired by Mammalian Vision?	126
53.	Why Vision? (ok, not for the "V")	127
54.	Deep Learning (bio)-inspired by Vision!	127
55.	AI Resurgence --› For Vision!	128
56.	Vision-Based Classification	128
57.	Real-Time Classification for Vision-Based Applications (case study Face Detection)	129
58.	Take advantage of opportunities!	129

PART 3 — 130

#	Title	Page
59.	Case Study I: Intelligent Motor Controller for UAVs	130
60.	Multi-Rotor UAVs	131
61.	Safety & Security	131
62.	Existing Dynamic Approaches	132
63.	Objectives (in collaboration with STMicroelectronics)	132
64.	So…what we did…	133
65.	Methodology	133
66.	Data Acquisition – Sensors & Logging	134
67.	Data Acquisition – Frequency and Format	134
68.	ANN Training Set Creation	135
69.	ANN Operation	135

70. Targeted Optimizations for Embedded ML	136
71. Code Optimizations	136
72. Data Optimizations	137
73. Experimental Evaluation	137
74. Experimental Platform	138
75. In-Field Evaluation	138
76. Experimental Results	139
77. Lessons learned?	139
78. But…What about when more data comes in the picture?	140
79. Case Study II: Vision for UAVs (Pedestrian & Car Detection)	140
80. Challenges!	141
81. Convolutional Neural Networks (CNN)	141
82. Small Deep-Neural Networks	142
83. Existing Approaches	142
84. Single Shot Detection– YOLO Approach	143
85. Single Shot Detection – YOLO Approach	143
86. Need efficient approach for object detectors	144
87. State-of-the-Art Challenges	144
88. Implementation Challenges	145
89. Dealing with computational cost of CNN	145
90. Data Collection & Training	146
91. Design Space Exploration of NN parameters	146
92. CNN optimization impact	147
93. DroNet Architecture	147
94. Performance Metrics	148
95. Single Combined Metric	148
96. Overall Performance for Vehicle Detection	149
97. Can we do better?	149
98. Performance and Accuracy Vs Resolution	150
99. Designing Neural Networks for Edge Intelligence (I)	150
100. Designing Neural Networks for Edge Intelligence (II)	151
101. Designing Neural Networks for Edge Intelligence (III)	152
102. Can We Get Inspired by Biology?	152
103. Tiling	153
104. Generation of Tiles	153
105. Memory Mechanism	154
106. Memory Mechanism (2)	154
107. Selective Tile Processing[1]	155
108. Attention Mechanism	155
109. Tiles Selection with Memory – TSM	156
110. Evaluation and Experimental Results	156
111. Tile Selection Metrics	157
112. Accuracy Vs CNN Input Size	157
113. Processing Time Vs CNN Input Size	158
114. Performance on Odroid XU4	158
115. Detection Results	159
116. Selective Tile Processing	159
117. Impact of Tiling, Memory & Attention	160
118. EdgeNet Framework Overview	160
119. EdgeNet Framework (I)	161
120. EdgeNet Framework (II)	161
121. EdgeNet Framework (III)	162
122. Initial Position Estimation	162
123. Tiling and CNN Selection (I)	163
124. Tiling and CNN Selection (II)	163
125. Optical flow (Lucas-Kanade[1] tracker)	164
126. Experimental Setup	165
127. Time-slot Selection	165
128. Time-slot Selection	166
129. Sensitivity	166
130. Average Processing Time	167
131. Average Power Consumption	167
132. DroNet in ACTION (I)	168
133. DroNet in ACTION (II)	168
134. DroNetV3 in action!	169
135. EdgeNet in action! (I)	169
136. EdgeNet in action! (II)	170
137. Observations	170

PART 4   171

138. Post Von-Neumann Computing	171
139. Technologies?	172
140. How about CMOS/EDA?	172
141. Conclusions	173
142. Thank You!	173

Machine Learning is nowadays embedded in several computing devices, consumer electronics and cyber-physical systems. Smart sensors are deployed everywhere, in applications such as wearables and perceptual computing devices, and intelligent algorithms power our connected world. These devices collect and aggregate volumes of data, and in doing so, they augment our society in multiple ways; from healthcare, to social networks, to consumer electronics and many more. To process these immense volumes of data, machine learning is emerging as the de facto analysis tool, that powers several aspects of our Big Data society. Applications spanning from infrastructure (smart cities, intelligent transportation systems, smart grids, to name a few), to social networks and content delivery, to e-commerce and smart factories, and emerging concepts such as self-driving cars and autonomous robots, are powered by machine learning technologies. These emerging systems require real-time inference and decision support; such scenarios therefore may use customized hardware accelerators, are typically bound by limited resources, and are restricted to limited connectivity and bandwidth. Thus, near-sensor computation and near-sensor intelligence are starting to emerge as necessities, in order to continue supporting the paradigm shift of our connected world. As such, traditional Von Neumann architectures are no longer sufficient and suitable, primarily because of limitations in both performance and energy efficiency caused especially by large amounts of data movement. To achieve this envisioned robustness, we need to re-focus on problems such as design, verification, architecture, scheduling and allocation policies, optimization, and many more, for determining the most efficient, secure and reliable way to implement these novel applications within a robust, resource-constrained system, which may or may not be connected. This lecture addresses topics in how we design this new systems, in terms of performance, energy and reliability, while operating with a limited number of resources, and possibly in the presence of harsh environments which may eliminate connectivity and pollute the input data. We also address challenges and opportunities using empirical case-studies involving drones in real-world applications.

HETEROGENEOUS CYBER PHYSICAL SYSTEMS OF SYSTEMS

1 1964: Stanley Kubrick's "Connected World"

- Before ARPANET
- Before the Internet
- Before Minski's & Papert's Vision Project (1966)

The notion of a real-time connected world has been a vision for quite some time, even if it's been classified as "science fiction" in the 60's.

2 KIOS Research and Innovation Center of Excellence - UCY

- ΚΟΙΟΣ – Inspired by Greek mythology
- "Kios" The Titan!
  - The *Titan of Intelligence*
  - ***Inquisitive mind and the questioning intellect***
- Est. 2008 ➔ Upgraded to CoE through a € 40 Million+ TEAMING Project (EU)
- Research on Intelligent Systems & Networks
  - Emphasis on Critical Infrastructure Systems
  - Real-Time, Decision & Support Mechanisms and Data Analytics

Sense ➔ Analyze & Decide ➔ React

In Greek mythology «Κοῖος», (pronounced Kee-os) the son of Uranos (Sky) and Gaia (Earth), was the Titan of "questioning intelligence" and the brother of Phoebe, the goddess of "answering intelligence."

Overview of the activities at KIOS

PART 1

Why should the Edge grow up?

---

3 — IoT: Complexity Trends and Challenges

- **Complexity** of IoT Systems is *exponentially increasing*
  - 10s of billions of connected heterogeneous devices

We live in an era where everything is now connected, and everyone has several "things" that constitute what is considered an edge device.

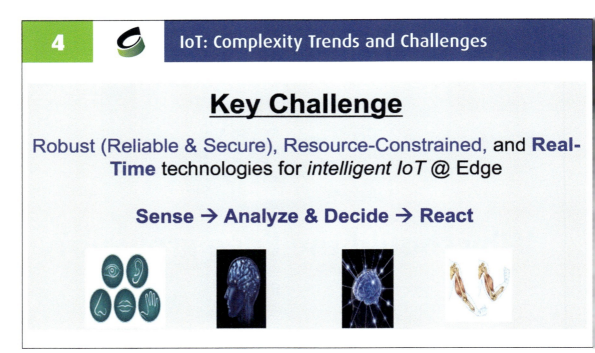

How do we make these devices smart? How can they "react"?

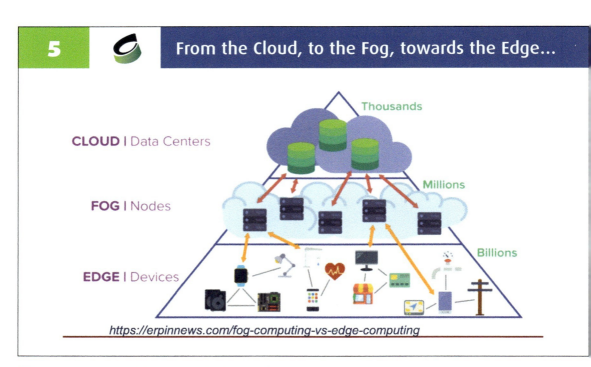

This gives rise to the Edge – Hub (Fog) - Cloud Paradigm

6. What does it mean in real-life IoT!

An Edge – Hub – Cloud Example.

7. Sensor Evolution

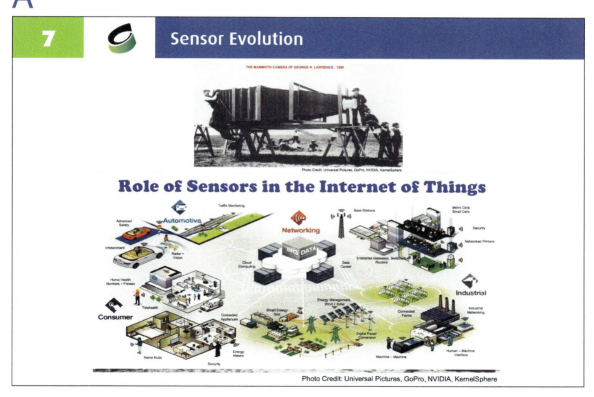

Sensors are what constitute the data collection of what we are reading in our modern world. They are cheap, can be purchased off-shelf, and can be connected right away on cheap boards, etc.

Modern IoT sensors are extremely mobile, very detailed and precise, and most importantly, CHEAP.

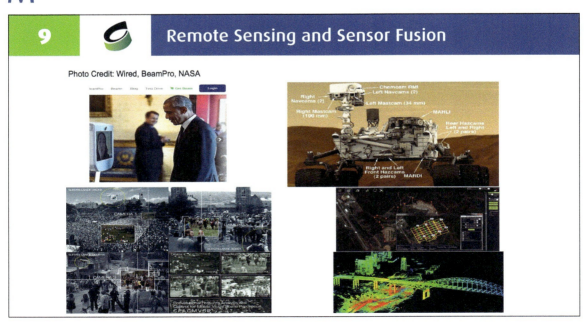

Sensors are also becoming mobile, and communication advancements enable sensors to reach remote and hard to sense areas. The sensors also cover wide data, data types, and data formats.

10 What about the Actuators?

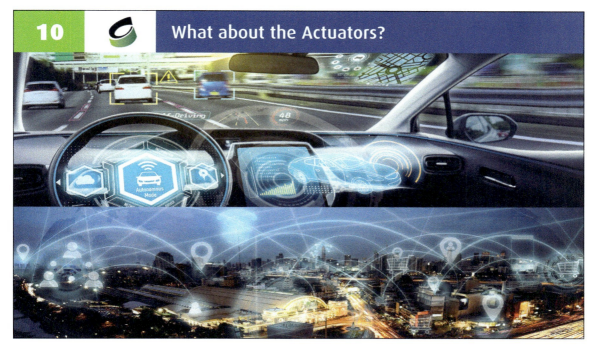

Another important driver in emerging cyberphysical systems are the actuators – the actuators can physically manipulate our physical world, and as a result, change the state of our surroundings, impact the environment and potentially alter our world.

11 So...

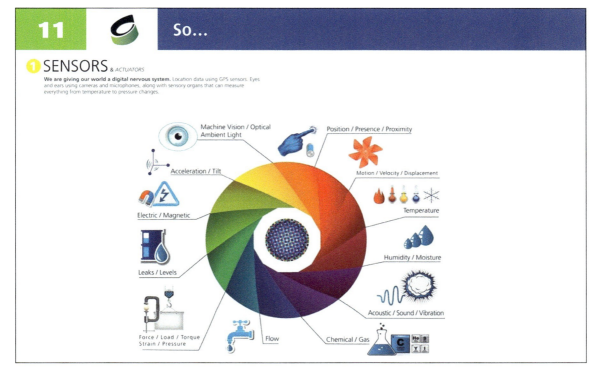

Examples of Sensors/Actuators.

But the Sensors/Actuators need a brain to process the data, derive context and put the context into action!

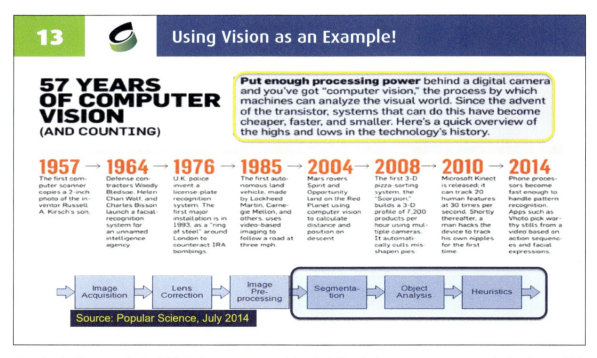

Such a brain example is visible in how computer vision is defined! The camera sensor reads the data, which is processed and enhanced. But it's the 'brain' that derives information and context and understands the scene. The brain is basically the computer processor architectures.

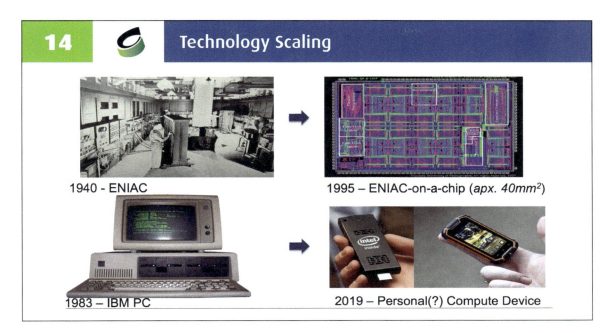

Technology scaling enabled us to "shrink" and miniaturize the brains.

But, we don't shrink things and solve all of our problems!

HETEROGENEOUS CYBER PHYSICAL SYSTEMS OF SYSTEMS

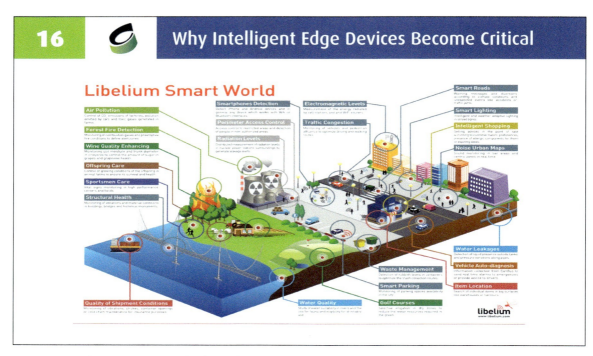

Integration of Computation with Physical Processes via Communication Network – Our world is entirely connected and sensing/processing and actuation are critical because they manipulate critical infrastructures!

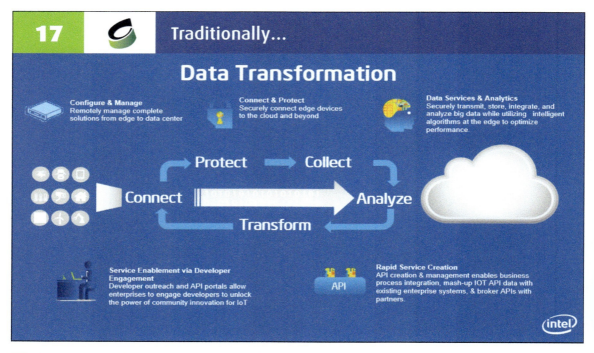

Traditionally, we shipped the data to a much more powerful brain, and received the brain's decision to perform the necessary action. We protected the shipment, we developed an infrastructure that enabled this paradigm.

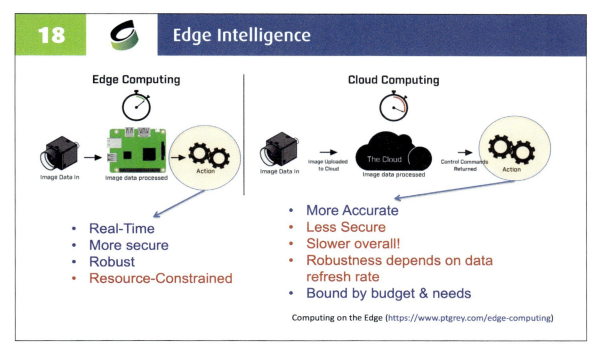

Why do we need to change this paradigm? Why do we need the edge to take its own decisions?

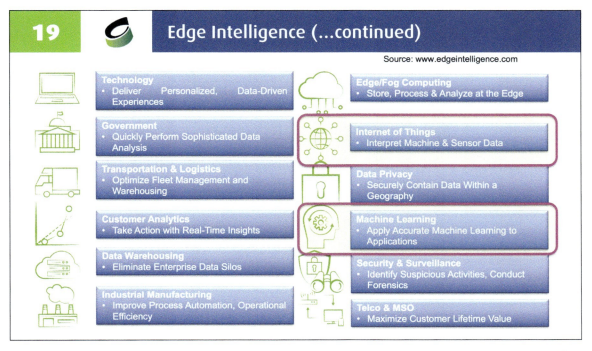

As a side note, edge intelligence encapsulates a wide range of areas. Including the buzzing ML and IoT.

20 — The Hype of Machine Learning

In 2016 when someone asked me "Are we really going to have 6+ MICRO papers a year on hardware support for AI/ML?". I said no, I think it's just this year and after there will likely be 3-5 per major conference. I was wrong; there were 12 papers on machine learning this year. Moreover, the trend does not appear to be slowing as it seems more than half the community is at least tangentially involved in a ML project.

While MICRO had a record number of ML papers (6 in 2016 and 4 in 2017), it didn't feel overwhelming and attendees didn't groan about "AI fatigue", at least not openly. This year there was a distinct focus on the practical problem of distributed training and new ML applications. Moreover, of the training papers, many proposed PIM based solutions or considered emerging memory technologies.

MICRO 2018 Summary
by Brandon Reagen on Nov 6, 2018

21 — Defining Artificial Intelligence

Artificial Intelligence
Mimics human behavior

Machine Learning
Uses advanced statistical algorithms to improve AI
- Regression
- Bayesian
- Clustering
- Decision Trees
- Vector Machines
- Neural Networks

Deep Learning (Neural Networks)
- Convolutional Neural Network
- Recurrent Neural Network
- Spiking Neural Network
- Capsule Neural Network....

- Artificial intelligence mimics human behavior
- Machine learning uses advanced statistical models to find patterns & results
- Deep learning is a specialized subset of machine learning using neural networks data to recognize patterns

However, What is ML? What's AI? What are all these buzzwords we keep hearing?

22 — What Makes it an AI SoC?

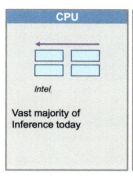
CPU — Intel
Vast majority of Inference today

GPU — NVIDIA
Vast majority of Training today
Enabled better-than-human-error capabilities

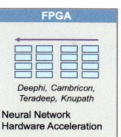
FPGA — Deephi, Cambricon, Teradeep, Knupath
Neural Network Hardware Acceleration

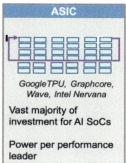
ASIC — GoogleTPU, Graphcore, Wave, Intel Nervana
Vast majority of investment for AI SoCs
Power per performance leader

- Most investment dedicated for CNN, RNN, some SNN
- Solutions include Software Development Kit (SDK) for mapping AI Graphs to hardware
- Most competitive include Neural Network Hardware Acceleration

What defines an SoC as an "AI chip"?

23 — AI's Insatiable Need for Compute, Memory, & Electricity

COMPUTE
- ResNeXt-101 - >30B operations
- Google's Voice Recognition >19B operations

MEMORY
- ResNet-152 - >60M parameters
- Google's Voice Recognition - >34M parameters

ELECTRICITY
"AI workloads could consume 80% of all compute cycles and 10% of global electricity use by 2030"

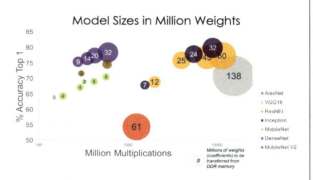

Model Sizes in Million Weights

Model Innovation is Increasing # of Weights & Multiplications Needed

AI is seemingly appearing everywhere, but at what cost? Compute? Memory? Electricity???

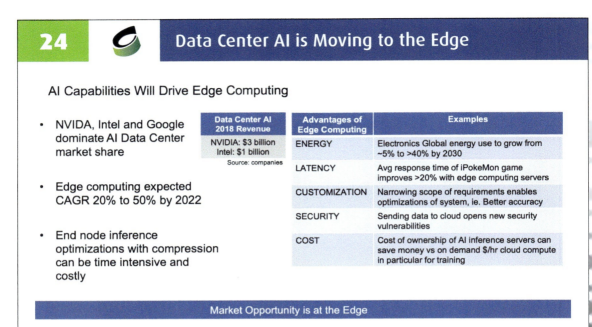

But...new technologies generate new opportunities, new opportunities generate new devices, and new devices generate new needs! Thus the paradigm shifts from the cloud to the edge!

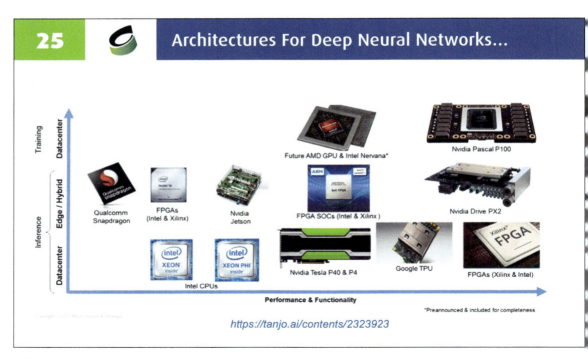

GPU: 200-250W / CPU: 90-100+W / Virtex: 18W, Altera 20-22 W / The race is ON!

26 — The Race for the Edge is ON!

Industrial examples.

27 — Google's Edge TPU and NVIDIA Jetson TX2

The Edge TPU is the little brother of the regular Tensor Processing Unit, which Google uses to power its own AI, and which is available for other customers to use via Google Cloud. Credit: Google

Industrial examples!

28 — Why Edge Intelligence Matters...

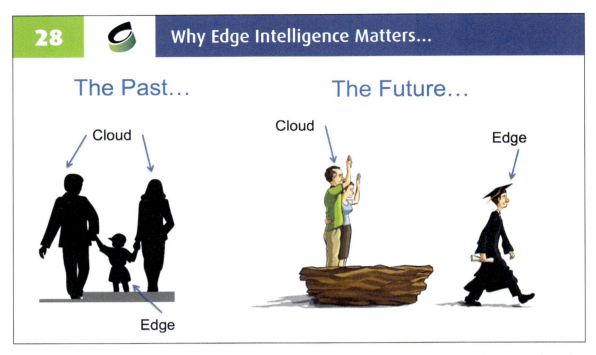

We can see it as a coming of age; if we keep relying on the cloud for decision support, eventually if the cloud is not available, the edge will collapse (brainless!) However, if we transfer the necessary intelligence towards the edge, the edge will continue operate in a secure, real-time and efficient manner!

29 — Deep Learning and Edge (IoT/SoC) Challenges – Nature vs. Nurture

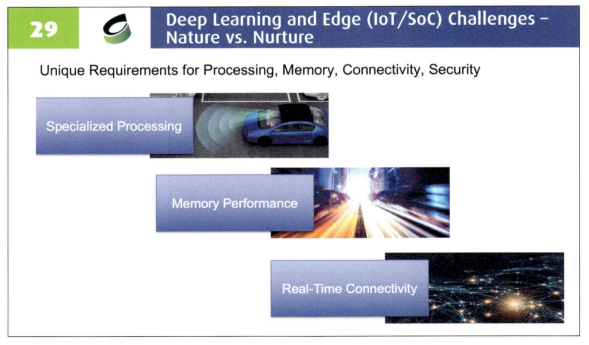

How do we take advantage of the nature of edge devices? How do we nurture the design of such devices?

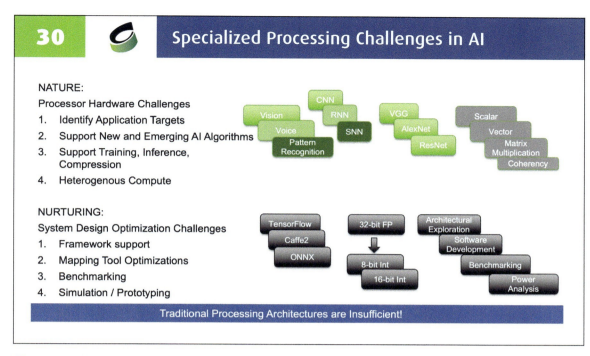

Processing Challenges.

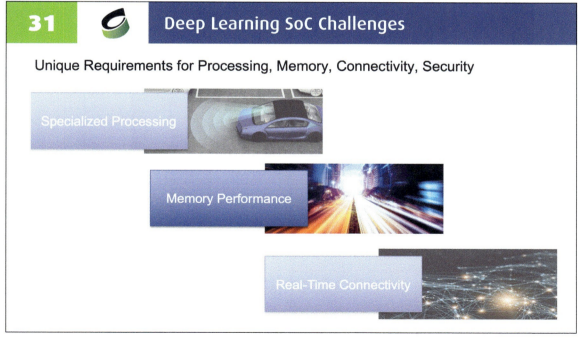

Memory

HETEROGENEOUS CYBER PHYSICAL SYSTEMS OF SYSTEMS

32 Memory Bandwidth at the Edge

Exercise in Processor Configuration & Mapping Tools

- Mobile & Auto use LPDDR
- Assumed Compression
- Available Techniques
  - **Quantization**: Converting from 32b FP to 8b/12b INT or FP
  - **Pruning**: Removing zero/near zero Coefficients
  - **Compression**: Reducing size of feature maps by removing statistical redundancy

Mobile AI Processors	Processing	Process Node	Memory Interface	Total Mem BW (GB/s)
Movidius Myriad 2 (gen 2)	12 VLIW	28nm	LPDDR3/2	7
Movidius Myriad X (gen 3)	16 VLIW + 2 NN	16nm	LPDDR4	9
Apple A11 Bionic	2 NN, 3 GPU	10nm	LPDDR4x	?
Apple A12 Bionic	8 NN, 4 GPU	7nm	LPDDR4x	?
Kirin 970	12 GPU, 1 NPU	10nm	LPDDR4x	30
Kirin 980	10 GPU, 2 NPU	7nm	LPDDR4x	34

Automotive Processors & Automation Level	Processing	Process Node	Memory Interface	Total Mem BW (GB/s)
Nvidia Xavier – Level 5	1 DLA, Vision, GPU, ISP, 8 ARM, 1 Video	12nm	x8 LPDDR4x	137
Tesla FSD Computer Chip – Level 5	2NN, 12 ARM, GPU, 1 Video	14nm	X4 LPDDR4	68
NXP S32V234 – Level 1	2 Image Processors, 1 ARM, GPU, ISP,	16nm	LPDDR3/2	4

Memory at the edge – examples and techniques to optimize.

33 Memory Performance Challenges in AI

NATURE:

IP Selection Addresses Memory Challenges (i.e. Synopsys EMLT)
1. Capacity (DDR5)
2. Bandwidth (HBM2e)
3. Power Consumption (LPDDR4x/5)

NURTURING:

But System Design Optimizations are required
1. Memory & Processing Co-Design
2. Large Array Yield Challenges (EMLT)
3. SRAM Customizations (EMLT)
4. Additional Test Vectors (SMS/SHS)
5. HBM2 Packaging Expertise & Support

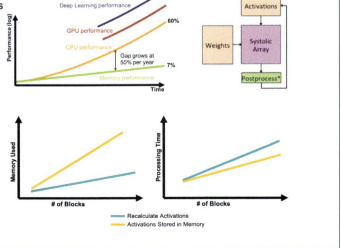

Memory Performance Challenges.

34 Deep Learning SoC Challenges

Unique Requirements for Processing, Memory, Connectivity, Security

What about the connectivity?

35 Real-Time Connectivity Challenges in AI

NATURE:
Great IP for AI Connectivity Challenges
1. Rapid 7nm Development
2. Cache Coherency
3. High Speed Chip to Chip
4. Latency

NURTURING:
System Design Optimizations
1. Area optimizations (i.e. DDR Hardening)
2. Time to Market (Subsystems)
3. Industry Expertise (Standards Expertise)
4. Early Software Development (Simulation/Prototyping)

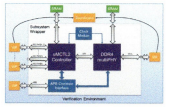

Connectivity challenges.

36 — Securing our little Edges!

Secure authentication, data encryption, key management, platform security & content protection

- AI Models
  - Expensive
  - Updates required
- AI applications use private data
  - Facial Recognition
  - Biometrics
- Integrity of the model:
  - Model corruption by nefarious agents
  - Corrupted models behave poorly

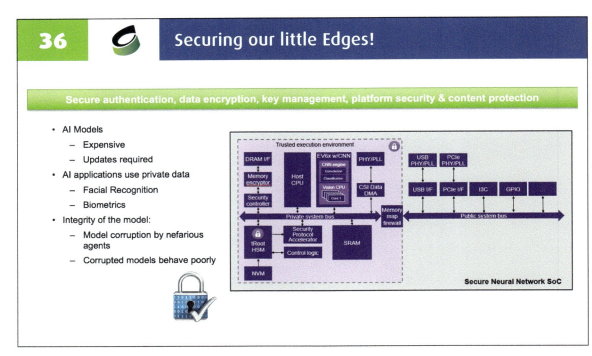

Security – a VERY important challenge. Just like we secure our kids, we need to secure our "edge" devices!

37 — Nurturing Amazing AI SoCs

- Expertise reduces risk, improves PPA, and improves time to market

- Industry Leading Tools enable more competitive designs

- Customizations optimize system performance

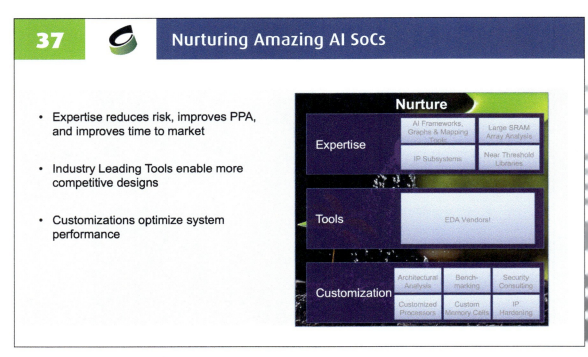

So...how do we help the "edge" to grow up???

PART 2

How can we help the Edge to leave the Cloud?

••• and leave the cloud?

38 Challenges and Opportunities of Edge Intelligence

- Shift computation towards the edge of the network
  - improve time-to-action and reduce latency
  - greater privacy and security
  - lower cost
- Challenges on the edge
  - limited compute resources (low-power devices)
  - low-memory
  - limited storage
- There is a need for optimization of the model and the structure based on the:
  - device H/W (embedded, FPGA, CPU, GPU) ,
  - algorithm,
  - task (e.g. pattern recognition, classification, decision support, control output, etc.)

There is in fact an opportunity to shift the computation toward the edge of the network to improve time-to-action and latency, to have greater privacy and security, and to minimize the cost by avoiding costly servers and communication infrastructures

In order to achieve that we need to face the challenges such as the limited compute resources by creating algorithm with low complexity. Minimize the memory and storage requirements in order to run those algorithms on the edge

There is a need of course, to optimize the models and the structure of machine learning algorithms, such as Convolutional Neural Networks based on the device that we want to run our algorithm, the algorithm and of course the computer vision task such as classification or detection

Challenges

Edge related constraints:

- Power Efficiency
- Cooling
- Battery
- Cost (overheads)
- Robustness
- Speed (latency)
- Physical Size...

- Implementation
- Prototyping
- Real-World Deployment
- Validation
- Evaluation
- ...

❏ Real-time operation (w.r.t. needs)
  → **Dedicated H/W?**
❏ Accuracy → **(Approximate Computing)**
❏ What **tradeoffs** can we exploit?
❏ **Hardware-friendly** solutions?

- Memory, Data-Flow...
- Pin-out limitations, buffering, bandwidth...
  - e.g. an HD frame in parallel... *1920x1080x32* pins!!!
- Arithmetic...
  - Bit-width precisions, accumulation, rounding, look-up tables, different number systems?

Constraints and Challenges.

Complexity of Neural Networks

**Huge Memory and Computational Requirements
(ResNET: 152 Layers, 11.3G MACs,
60M weights)**

❏ Different DNN architectures and their resource requirements

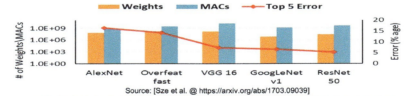

Source: [Sze et al. @ https://arxiv.org/abs/1703.09039]

Framework	fps (NVIDIA Jetson TX1)	IOU/mAP.
Fast YOLO	17.85 [1]	52.7 [2]
O-YOLOv2	11.8 [1]	65.1 [1]
YOLOv2	5.4 [1]	67.2 [1]

Sources:
[1]: Shafiee et al. @ arXiv:1709.05943 (2017)
[2]: Wei@ECCV1'6

Neural networks, the de-facto tool for intelligent systems today, is a complex, demanding and power-hungry paradigm!

41 Designing Neural Networks for Edge Intelligence

- **Deep Neural Networks**
  - decision making and data analytics
  - **high** computation and memory requirements
- **Current Trends**
  - **Design & Train a CNN to achieve max. accuracy & THEN Optimize for Edge:**
    - reduce the bit-width precision of the network parameters and the processed data
    - use binary values
    - use compression techniques with quantization and pruning to reduce the network demand for memory and storage
      - [1]XNOR-Net
      - [2]Deep Compression

[1]M. Rastegari, V. Ordonez, J. Redmon, and A. Farhadi, "Xnor-net:Imagenet classification using binary convolutional neural networks."
[2]Han, H. Mao, and W. J. Dally, "Deep compression: Compressing deep neural network with pruning, trained quantization and Huffman coding"

Neural networks are increasingly being used in many applications to provide decision making and data analytics (CNNs for computer vision tasks).

Artificial Intelligence (AI) is currently facilitated using machine learning algorithms such as **Deep Neural Networks(DNN)** which demands high computation and memory requirements.

Therefore, the computation was pushed from device to remote computing infrastructure.

Current trends for designing Neural network for the edge are trying to reduce both memory and computational requirement by reducing the bit-width precision of both the network parameters and processed data.

Another techniques is to use binary values for the representation of the weights in order to avoid floating point operations which are high computational demanding.

Lastly compression techniques with quantization and pruning of the network has show significant impact of both memory and storage. Some examples are the Xnor-Net and the deep compression network.

42 Energy-Efficient Deep Learning

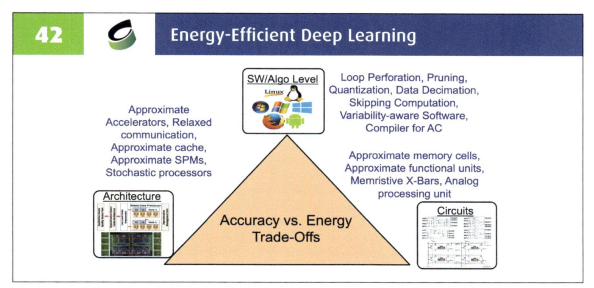

Need to look at it at ALL levels (Architectures, S/W and Algorithms, and Circuits)

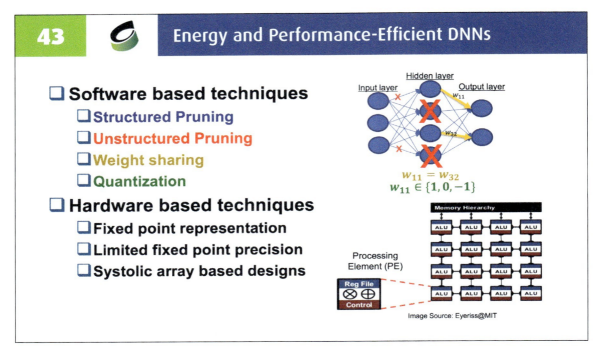

What we can do so far? What is the current practice?

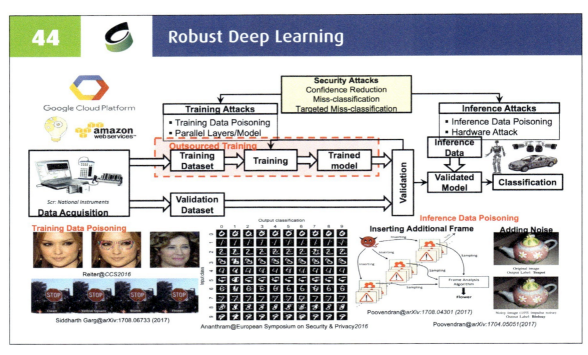

Robustness is a very demanding issue!

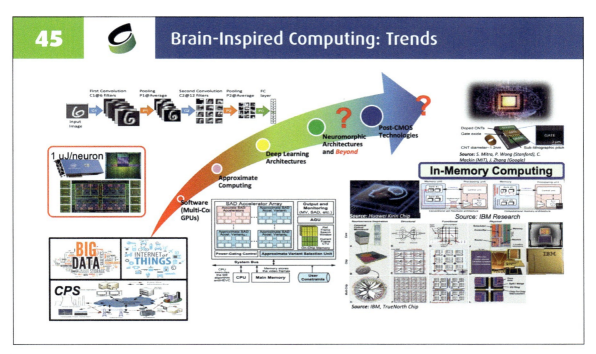

What does the future look like?

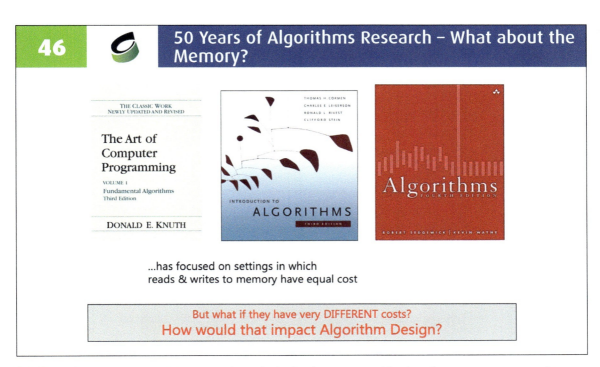

What about the memory? We haven't yet looked in how to possibly alter the computation paradigms to focus on the memory!

47 — Emerging Memory Technologies

We are on the cusp of the new non-volatile storage class memory which is also energy efficient and enables in-memory computing!

48 — Emerging Memory Technologies

Motivation:
- DRAM (today's main memory) is volatile
- DRAM energy cost is significant (~35% of DC energy)
- DRAM density (bits/area) is limited

Promising candidates:
- **Phase-Change Memory (PCM)**
- Spin-Torque Transfer Magnetic RAM (STT-RAM)
- **Memristor-based Resistive RAM (ReRAM)**
- Conductive-bridging RAM (CBRAM)

Key properties:
- **Persistent**, significantly lower energy, can be higher density
- Read latencies approaching DRAM, byte-addressable

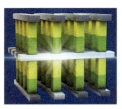

3D XPoint

New memory technology examples. Focus on PCM and ReRAM.

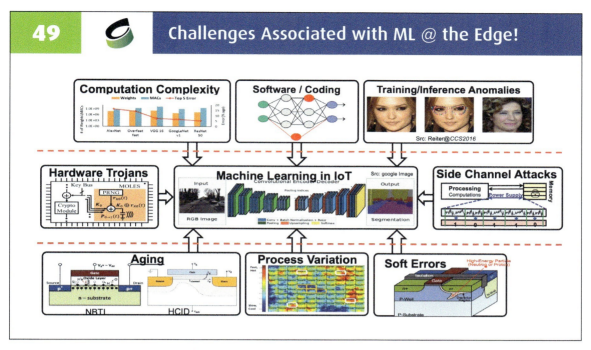

So with that said…what are the challenges? This figure summarizes what we discussed so far!

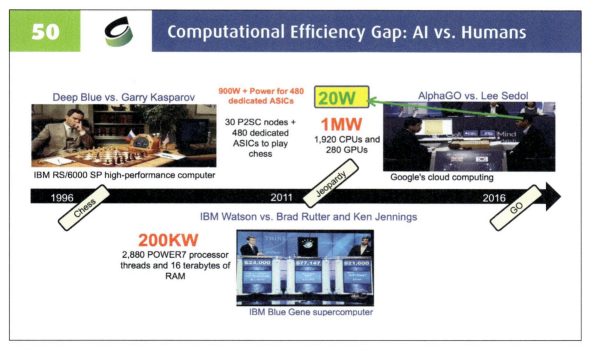

How can we overcome these challenges? Where does AI come from? Recall that the objective is to MIMIC human behavior! So why don't' we start by studying how human brains or mammalian brains solve problems!

51 — Why Brain-Inspired Computing...

Power-Efficiency
- *20W* consumption
- *Always* ON

Learning Capability
- Supervised/Unsupervised
- Object identification
- Language understanding

Fault-Tolerance
- Noisy Inputs
- Neurons may die

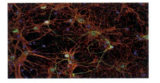

Our brain is an amazing tool!

52 — ...for Big Data? Inspired by Mammalian Vision?

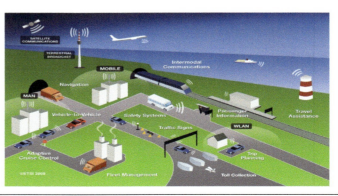

In fact, we can get inspired by mammalian vision primarily because we have significant expertise now with computer vision.

53 — Why Vision? (ok, not for the "V")

BIG DATA

According to Merriam-Webster:
- an accumulation of data that is too large and complex for processing by traditional **database management tools**

According to Google:
- extremely large data sets that may be analyzed computationally to reveal **patterns, trends, and associations**, especially relating to human behavior and interactions.

INTERESTING VISION FACTS:
- Two thirds of the brain electrical activity (2/3 billion firings /s) when eyes open.
- 50% of our neural tissue **directly** (or indirectly) related to vision
  Qnspac8P, Q, Dgrm* L cspm_l _mk gr* / 735
- More neurons dedicated to vision than all four senses combined
- Olfactory cortex losing ground to visual cortex
  (i.e. vision is "eating" our smell!)
  Qnspac8Hrf l K cbg _* @_g Psjcq* 0. / 3

Computer Vision has been around for a while, but also mammalian vision is a superb data analysis tool!

54 — Deep Learning (bio)-inspired by Vision!

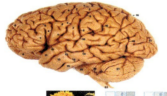

- **Our Brain**
- >10^{16} **complex operations / second** (10 Petaflops!!!)
- **10-15 watts!!!**
- 1.5 kg

Nature achieves **efficient and reliable** computation based on **fuzzy input** data in an **uncontrolled** environment

Ellis et al. "human cross-sectional anatomy" 1991, Ed. Butterworth

"K computer" (RIKEN, Japan)

8.162 petaflops
9.89 MW

http://www.nsc.riken.jp

In fact, computer vision is the driver for modern artificial intelligence research and deep neural nets!

55 — AI Resurgence --> For Vision!

The first dedicated neural chips were based on vision applications. An example includes Intel Movidius neurocomputing stick.

56 — Vision-Based Classification

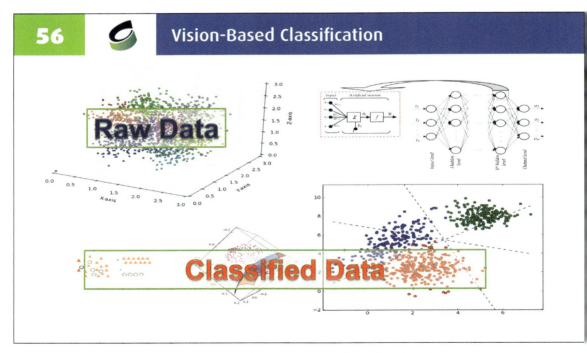

In fact, computer vision's primary goal is to classify raw data . Thus, algorithms based on computer vision can be derived that operate on other sensor data (i.e. time-series data, etc.)

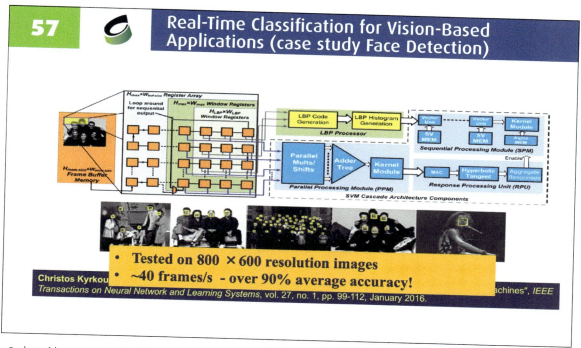

A lot of low-power classification works in existence, so it's easier to start from somewhere we know!

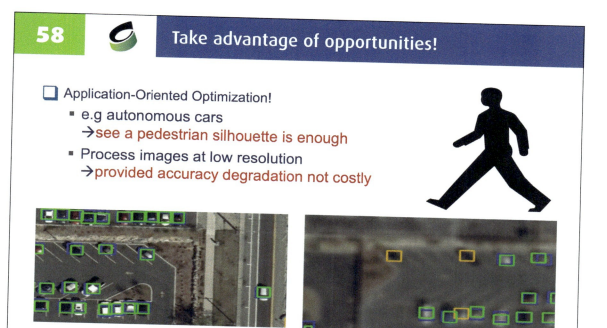

But, much like humans, let's start by taking advantage of opportunities, derived as to how we, humans that is, would do things!

PART 3

A couple of Grown-Up Edges...

Lets take a look at a couple of example case-studies with different requirements and different constraints!

59 — Case Study I: Intelligent Motor Controller for UAVs

- Rotors controlled by remote-control (fly-by-wire) and inertial on-drone sensors for feedback

- Can we build an intelligent controller which will read-and-react taking unforeseen situations into consideration?

G. Michael, N. Efstathiou, K. Mantis, T. Theocharides and D. Pau, "Intelligent embedded and real-time ANN-based motor control for multi-rotor unmanned aircraft systems," *2017 IFIP/IEEE International Conference on Very Large Scale Integration (VLSI-SoC)*, Abu Dhabi, 2017, pp. 1-6. doi: 10.1109/VLSI-SoC.2017.8203456

Lets start with the first case study. This case involves the design of an intelligent motor controller for recognizing unforeseen disturbance on a drone.

60 Multi-Rotor UAVs

- Commercial multi-rotor UAVs emerging as the "hot" gadget of this decade
  - Various application domains
  - Huge market ratio

- But...what is the catch?
  - Easy to fly (my 5 year old daughter flies one)
  - Hard to follow the rules, and the safety procedures

- Some Universal Ground Rules:
  - Drone pilots must follow the aviation rules
  - Drone pilots must ensure that the equipment is correctly, and properly parameterized (checklists)
  - Drone pilots must appropriately use fail-safe mechanisms

Multirotor Drones are available everywhere. However, while this is a good thing, it's also a bad thing. They are found in applications where there are many many dangers, so we need some sort of on-board intelligence to recognize unforeseen environmental disturbance.

61 Safety & Security

- Flight controller provides fail-safe mechanism
  - Mostly based on GPS modules
  - Failsafe mechanisms for returning home, landing autonomously, etc.
  - Obstacle avoidance and safe-landing mechanisms also available
  - Emergency Mode (i.e. panic landing, rotor turn-off, etc.)
- Failsafe mechanisms insufficient for the extend of the market!
- But...what did we miss?
  - Technical problems (motors fail, GPS fail, etc.)
  - Unforeseen flight incidents (turbulent winds, inexperienced pilots, etc.)
  - **INCREASING COSTS!**
    - **Insurance, Training, Legal & Business Framework**
    - **Cost of Drones – More Reliability = $$$**

Existing on-board solutions not sufficient.

62 — Existing Dynamic Approaches

Traditional control:
- PID control
  - Most common solution due to its simplicity and effectiveness
  - Time-consuming and laborious process for real-time results
- H-infinity Control
- Predictive Control
- Neural Networks
  - Create non-linear mapping from input to outputs and capture the UAV flight dynamics
  - Robust controller
  - One solution is an adaptive neural network controller based on CMAC (Cerebellar Model Articulation Controller)
    - Power Hungry, Memory-Hungry!

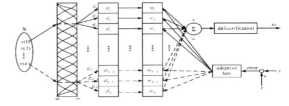

Current Approaches lie in traditional control, but for ANN-based control, traditional designs are not suitable.

63 — Objectives (in collaboration with STMicroelectronics)

- Aim to develop a light-weight ANN-based controller constrained by the physical dimensions, power consumption and real-time computation, which will be capable of
  - Real-time recognition & reaction (*throttle, pitch, roll, yaw*)
  - Fit on a battery-powered embedded processing platform
  - Run using onboard sensing data (rely on the drone rather than the user or the network)

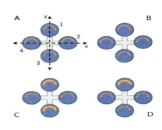

We aim to develop a light-weight ANN-based controller constrained by the physical dimensions, power consumption and real-time computation, which will be capable of
- Real-time recognition & reaction (*throttle, pitch, roll, yaw*)
- Fit on a battery-powered embedded processing platform
- Run using onboard sensing data (rely on the drone rather than the user or the network)

64 So...what we did...

- **Data Collection & Training Data Set Construction**
  - Attach a flight controller to a commercial drone, induce various unwanted scenarios, record sensor data
- **ANN development**
  - Use sensor data to form a discrete time-series input
  - Use expected vs. unexpected motion
    - Comparison by reading remote control signals concurrently
- **Reliance on self-collected *and available* data!**
  - Sensor data are collected only from drone
    - No GPS, no externally collected data
- **Embedded ANN Deployment & Optimization**
  - Use approaches such as pruning, quantization, etc. to reduce ANN footprint and complexity for easy deployment on a common embedded board
- **Experimental Validation**
  - Validate controller in a real-world scenario using a drone!

65 Methodology

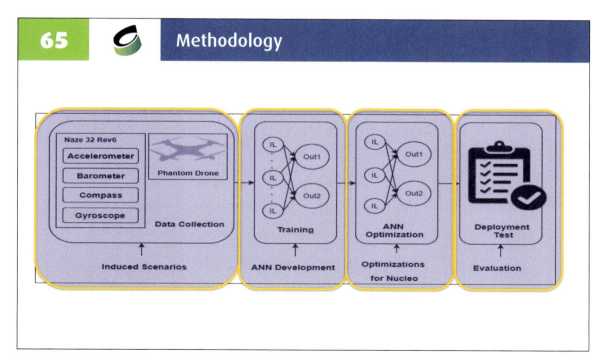

We first Identified the sensor data and the types of data. Then we developed our ANN model, and performed optimization. We then evaluated using in-field simulations.

66 Data Acquisition – Sensors & Logging

- On-Board Sensor Data
  - Accelerometer, Gyro, Compass, Barometer (Altimeter)
- Integrated setup for data collection (training data):
  - A single board computer (Raspberry Pi 3 / Odroid XU4) integrated with a flight controller (Naze 32 Rev6)
  - The above deployed to a commercial UAV (DJI Phantom 2)
- Setup powered by a USB power bank
  - Naze 32 power-up through the Raspberry pi, and the Raspberry pi power-up through USB power bank

67 Data Acquisition – Frequency and Format

- Naze 32 sensors → Accelerometer, Magnetometer (Compass), Barometer (Altitude) and Gyroscope
  - Sampling intervals set at 20/second.
  - ANN processing time per sampling window set at 2 seconds (e.g. 40 discrete values / time series format per sensor reading)

Data	Type	Details
accx	INT 16	unit: it depends on ACC sensor and is based on ACC_1G definition
accy	INT 16	unit: it depends on ACC sensor and is based on ACC_1G definition
accz	INT 16	unit: it depends on ACC sensor and is based on ACC_1G definition
gyrx	INT 16	unit: it depends on GYRO sensor. For MPU6050, 1 unit = 1/4.096 deg/s
gyry	INT 16	unit: it depends on GYRO sensor. For MPU6050, 1 unit = 1/4.096 deg/s
gyrz	INT 16	unit: it depends on GYRO sensor. For MPU6050, 1 unit = 1/4.096 deg/s
magx	INT 16	unit: it depends on MAG sensor.
magy	INT 16	unit: it depends on MAG sensor.
magz	INT 16	unit: it depends on MAG sensor.
EstAlt	INT 32	cm

68 ANN Training Set Creation

Develop & Train an ANN
(MLP, CNN, RNN, etc.)
- **CNN and RNN Unnecessary!**
- ANN takes as input the four sensor data:
  - Barometer → **1 int. value**
  - Gyroscope → **3 int. values**
  - Magnetometer → **3 int. values**
  - Accelerometer → **3 int. values**

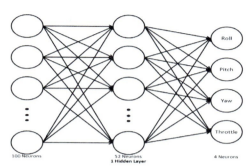

Multi-layer perceptron, initialized with 10*20 (200) nodes
Resulted in 102 hidden nodes
Four output neurons (roll, pitch, yaw, and throttle)

We first identified the targeted ML algorithm. From some initial experiments we realized that we did not need to make it too complicated!

69 ANN Operation

INFERENCE:
- The ANN receives data from the sensors during a sampling window of *n* intervals
  (sampling dependent on sensing & ANN processing speed)
- ANN *recognizes* unwanted behavior
  - returns adjustments of the pitch, yaw, roll and throttle signals at the end of every sampling window
- IMPLEMENTATION
  - STM Nucleo-144 (ANN)
  - NAZE32 Flight Controller (Sensor Data)
  - USB Power Bank for power
  - *Rasberry Pi 3 (Logging & Debugging)*

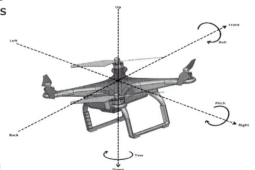

This is a description of how the ANN operates.

70 Targeted Optimizations for Embedded ML

- Embedded board constrained by limited resources
- Target Real-time performance,
  - Both fast and power-efficient ANN
  - Lightweight in terms of size
  - Performed certain optimizations such as
    - reduce execution time
    - lower the power consumption
    - minimize the memory footprint of the ANN
    - reducing the memory footprint enabled us to increase the range of neuron instances that could be created in parallel

NO NEED FOR DEEP NEURAL NETWORK!

The conclusion is that we do NOT need too much! Just because DNNs and DL are buzzing everywhere, it does not make them suitable for every application!

71 Code Optimizations

- Weights directly hard-coded as an array
  - avoid linking memory-intensive libraries
- Circular buffer to store the data
  - avoid usage of dynamic memory allocation
- The targeted STM compiler (GCC) configured for performance
- Raw data normalized and *tupled* once every timing window (i.e. every *dt*);
  - data stored in its new form and reused over its cycle lifespan.

72 Data Optimizations

❏ Introduced an **orientation filter** which uses a **quaternion representation**, in order to fuse accelerometer, magnetometer and gyroscope data into a representation of the drone in a *three-dimensional space* ➔ *data reduction*
  - Reduction of input sensory data per tuple from 10 to 5
  - The number of input neurons was reduced from 200 down to 100, the number of necessary hidden nodes to 52 and the number of total weights to less than half of the original

❏ **Accuracy impact (approximately 3-5 %)**
  - *Benefits are extremely important*

Sometimes data representation is very important.

73 Experimental Evaluation

❏ Two-Phase evaluation
- First phase includes evaluation tests in the lab
  - without propellers (using simulator)
  - Evaluated Performance & Accuracy

❏ Second Phase involves experimental evaluation outside!
  - Compared the accuracy of the ANN when running induced motion trajectories (some were also part of the training set, such as drifting and tilting)
  - Monitored output of the ANN through a Raspberry Pi.
- All flights were video recorded in order to be able to verify the ANN result in the lab

1. Raspberry Pi
2. STM Nucleo Board
3. USB Power Bank
4. NAZE32

Need in-field experimental evaluation!

74 Experimental Platform

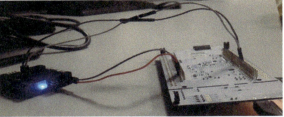

75 In-Field Evaluation

Induced various scenarios
- *Attached fishing line to drone!*

Disclaimer:
No students, drones or the environment were harmed during this process!!!

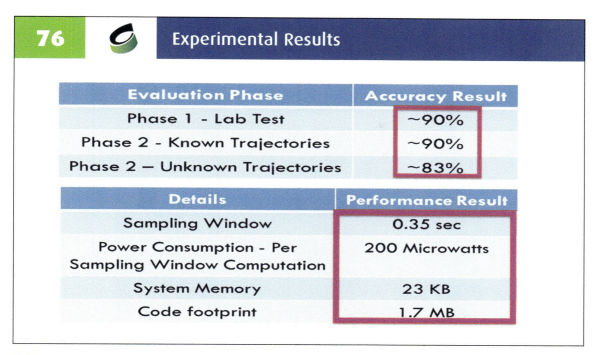

Very good accuracy and very good results.

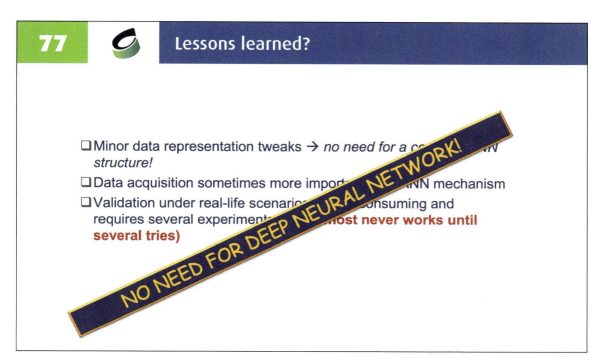

The conclusion is that we do NOT need too much! Just because DNNs and DL are buzzing everywhere, it does not make them suitable for every application!

78 But...What about when more data comes in the picture?

- Drone's inertial sensors sufficient for situational awareness
- What if we utilize its vision sensors?
- What if we consider human intentions (or **non**-intentions!)
- **Eventually will have to migrate to more complex AI**

What happens though when the amount of data increases?

79 Case Study II: Vision for UAVs (Pedestrian & Car Detection)

- **UAV payloads include an array of visual sensors (IR, LiDar, etc.)**
- **Potential CIS applications**
  - Emergency Response, Traffic Monitoring, Surveillance
- **Detecting objects is a key step for many applications**
  - Boost remote sensing and situational awareness

Lets find out via a second case which involves an aerial (Drone based) visual object detection system.

80 Challenges!

- UAVs equipped with high resolution cameras (<4k) and fly in high altitudes,
  - *Vision on the edge becomes challenging as the image resolution increases*
- Larger networks incur higher computational cost
- Reducing the image size effectively reduces the object resolution and detail → Difficult to detect smaller objects such as pedestrians

The challenges we have to face is the LARGE volume of data stemming from 4K resolution sensors, and of course a moving camera in high altitudes and with very large occlusion probabilities.

81 Convolutional Neural Networks (CNN)

- Convolutional Neural Networks (CNNs) are biologically inspired hierarchical models that can be trained to perform a variety of image detection, recognition and segmentation tasks.
- Single-Shot Detectors: Pose detection as a regression problem which takes an input image and learns the class probabilities and bounding box coordinates
  - Aim to avoid the performance bottlenecks of the region- and window- based systems.

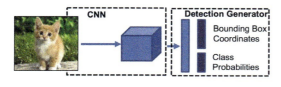
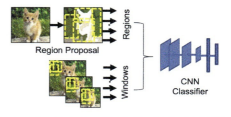

Typically, when dealing with visual object detection, convolutional neural networks are unparalleled. During the last few years especially the advancements are tremendous both in terms of accuracy and also in terms of performance.

82   Small Deep-Neural Networks

- Small DNNs are more deployable on embedded processors
  - Computation and even more so memory are at a premium
  - Storing the model on chip saves on power, and improves performance

- Faster to go through training iterations
  - more easily updatable over-the-air (OTA)

- Small DNNs are more power efficient
  - Less off-chip memory accesses which consumes order of magnitudes more power.

- Small DNNs permit for multiple vision tasks to run on the same platform e.g., object detection

Han et al, "Deep Compression: Compressing Deep Neural Networks with Pruning trained quantization, and Huffman Coding", ICL 2016

Advancements have also occurred across resource-constrained embedded devices.

83   Existing Approaches

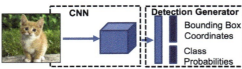
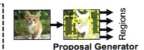

Single Shot Detectors
- Yolo(2015)
- Yolo9000 (v2)(2016)
- SSD(2016)

- Faster but still demanding for edge
  - size > 5 GB of working memory
  - 6.97 billion operations
- Capture global as well as local context
- Lower Memory Footprint
- **Less Accurate**

Region-Proposal based Detectors
- R-CNN(2014)
- Fast R-CNN(2015)
- Faster R-CNN(2015)

- Typically more accurate
- Slower – require multiple passes of numerous regions

Huang, Jonathan, et al. "Speed/accuracy trade-offs for modern convolutional object detectors." *IEEE CVPR*. 2017.

Combine ideas from both single-shot detectors and region-proposal detectors.

84. Single Shot Detection– YOLO Approach

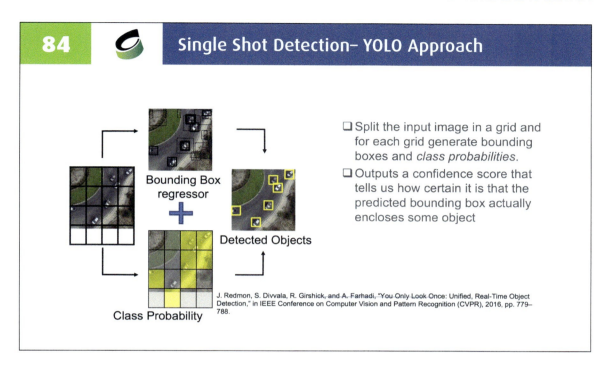

- Split the input image in a grid and for each grid generate bounding boxes and *class probabilities*.
- Outputs a confidence score that tells us how certain it is that the predicted bounding box actually encloses some object

J. Redmon, S. Divvala, R. Girshick, and A. Farhadi, "You Only Look Once: Unified, Real-Time Object Detection," in IEEE Conference on Computer Vision and Pattern Recognition (CVPR), 2016, pp. 779–788.

85. Single Shot Detection – YOLO Approach

- The confidence score class prediction is combined into one final score that tells us the probability that this bounding box contains a specific type of object.
- Predicts B bounding boxes, confidence for those boxes, and C class probabilities
- Trained end-to-end using Gradient Descent
- Major Advantage is **Real-Time Performance**

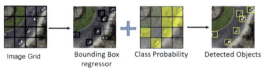

tinyYOLO and YOLO networks

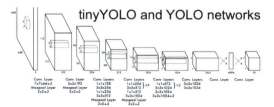

J. Redmon and A. Farhadi. 2017. YOLO9000: Better, Faster, Stronger. In 2017 IEEE Conference on Computer Vision and Pattern Recognition (CVPR). 6517–6525.

86 Need efficient approach for object detectors

- Optimization beyond the CNN architecture
  - Can process higher resolution images without significant computational cost → Make smaller objects detectable
  - Discard information and avoid unnecessary computations
  - Avoid reducing the image resolution and distorting the objects

Obviously accuracy is always the desired result. But, we can optimize as much as we can, keeping a minimum acceptable accuracy as a constraint.

87 State-of-the-Art Challenges

- **Existing pretrained networks for object detection:**
  - CNN not suited for aerial view of objects ← lack of data & limited effort!
  - target mostly GPU platforms (power hungry, cost!)
  - target processing of aerial view images only for satellite applications so far
  - (not) suitable for edge *just starting*!
  - Need to eliminate redundant information – unnecessary objects (e.g. bicycle)

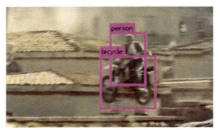

These are the challenges for the specific application.

88 Implementation Challenges

- High-end computing infrastructure not available
  - especially in emergency response scenarios
  - remote areas may not have access to infrastructure
- Cloud Implementation
  - high latency (mostly communication!)
  - connectivity and robustness
  - security risk (data privacy, etc.)
- Fog Implementation (ground station)
  - may address latency & power
  - connectivity/robustness Issues
  - some security risks!
- Local, on-board processing is necessary but…
  - Computer Vision Algorithms are Expensive
    - Limited memory footprint, compute resources
  - "stealing" from battery limits flight time!
  - adding payload weight limits flight time!

Processing high resolution images on resource-constraint systems such as UAVs is an open problem

89 Dealing with computational cost of CNN

- What is a suitable CNN architecture that can be used to identify objects from on-board the UAV?

- optimize CNN algorithms for detecting vehicles and pedestrians
- efficient and suitable for real-time UAV applications
- running on the edge (e.g. an embedded hardware platform).
- parameter space and design-space exploration in the design of convolutional neural networks and identify an efficient architecture

We needed a way to handle the computational cost of the CNN algorithm.

What is actually a suitable CNN architecture that can be used to identify object from on-board the UAV?

In order to find that our we adapt existing CNN algorithms for detecting both vehicles and pedestrians and we attempt to make them efficient and suitable for real-time UAV applications on the edge.

Our methodology basically explore the parameter space in the design of the CNN and identify and efficient architecture

90 — Data Collection & Training

- **Images were collected using a variety of methods**
  - cropping of satellite images
  - retrieving images from the world-wide-web
  - collecting urban traffic video footage from a UAV
- **All vehicles in the images were manually annotated in order to obtain the ground truth**
- **Training/Inference**
  - DarkNet is an open source neural network framework written in C and CUDA.
  - Fast, supports CPU and GPU computation

91 — Design Space Exploration of NN parameters

- Explore the effect of different parameters
  - Number and types of layers
  - Filter Number and Size
  - Input Image size
- A comprehensive quantitative evaluation of CNN-based vehicle/pedestrian detectors
  - Tiny-Yolo (Fast on GPU)
  - TinyYoloNet
  - SmallYoloV3
  - DroNet[1] (Fast on CPU)

 SmallYoloV3   DroNet   tinyYoloNet   tinyYoloVoc

[1] Christos Kyrkou, George Plastiras, Stylianos Venieris, Theocharis Theocharides, Christos-Savvas Bouganis, "DroNet: Efficient Convolutional Neural Network Detector for Real-Time UAV Applications", In proceedings of International Conference on Design Automation and Test in Europe (DATE), March 2018.

In order to find out the best CNN-based structure we explore the effects of numbers and types of layers, the filter number and size and the input of the CNN

We created different networks and evaluated each one using a score function which shows which design its better for our application.

92 — CNN optimization impact

- **Memory Footprint / Performance**
  - Reduce the number of layers
  - Reduce number of filters
  - Reduce filter size
  - Increase stride
- **Object Size**
  - Affects the accuracy
  - Choose input based on object
- **Input Size**
  - Increase for better accuracy
  - Reduce for better performance

[1]https://pjreddie.com/darknet/yolo/?utm_source=next.36kr.com

CNN	CNN Size
Tiny-Yolo[1]	63.1MB
DroNet	283KB

Lets start with the reduction of memory footprint and the increase of the performance.

First we analyze the impact of each parameter on the performance using the Billion floating operations metric.

We noticed that in order to reduce the floating operations we needed to reduce the number of layers, the number of filter, the filter size and increase the stride.

All of this while trying to maintain the accuracy of the detector high. This reduction of the parameters also has a significant impact on the CNN size.

On the table we can see that compared to the tiny-yolo which is a CNN-based detector which is really fast on gpu, we managed to reduce the size from 63.1mb to 283kb

93 — DroNet Architecture

- Suitable for fast inference on UAVs
- Make use of 3x3 filters
- Progressively reduce the feature maps size by a factor of 2
- Increase number of filters but use 1x1 convolutions at deeper layers
- Higher number of filters at the beginning to capture more image features

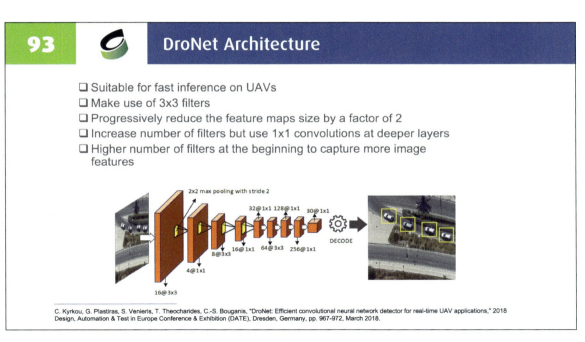

C. Kyrkou, G. Plastiras, S. Venieris, T. Theocharides, C.-S. Bouganis, "DroNet: Efficient convolutional neural network detector for real-time UAV applications," 2018 Design, Automation & Test in Europe Conference & Exhibition (DATE), Dresden, Germany, pp. 967-972, March 2018.

Towards this directions we have proposed Dronet for use in UAV applications. Our initial work focused on vehicle detection.

94 Performance Metrics

To evaluate the performance of each model, we employ the following four metrics:

$$\text{Sensitivity}: \frac{T_{pos}}{T_{pos} + F_{neg}}$$

$$\text{Precision}: \frac{T_{pos}}{T_{pos} + F_{pos}}$$

$$\text{IoU}: \frac{\text{Area of Overlap}}{\text{Area of Union}}$$

$$\text{FPS}: \frac{1}{t_{proc\_per\_frame}}$$

At the same time, we need to define performance metrics which are suitable for the application.

95 Single Combined Metric

- To capture the overall performance of each detector, we define a composite linear combination metric

$$Score(w) = w_1 \times FPS + w_2 \times IoU + w_3 \times Sensitivity + w_4 \times Precision \qquad \sum_{i=1}^{4} w_i = 1$$

- Parameterize the score with respect to a vector of weights, where each weight captures the application-level importance of each metric.
- Prioritized FPS with a weight of 0.4 over the other three accuracy-related metrics, which were equally weighted with 0.2.

It is important to combine the performance metrics for a specific application in order to find the one that application requirements.

In our case we prioritized FPS for a real-time performance with a weight of 0.4 and equally weight with 0.2 all the other metrics. However, this is an adjustable optimization constraint that can be adopted application's needs.

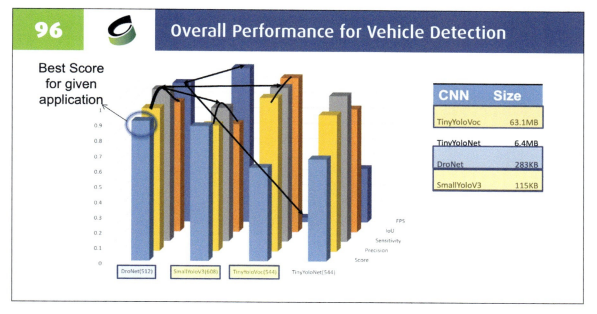

Example for vehicle detection. Based on how we targeted the optimization of our design approach, the result we obtained was better.

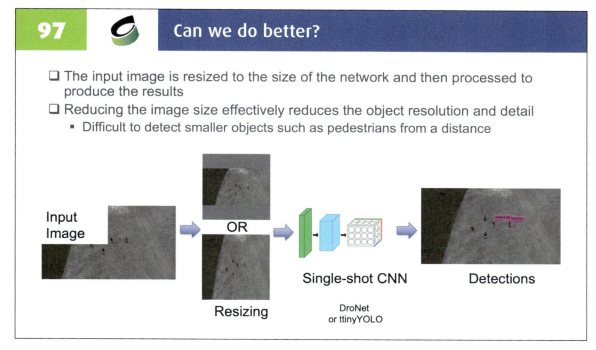

The standard approach when using either the tinyYOLO or DroNet networks is to resize the input image to the size of the network process it and get back the results. For smaller objects from a distance the detection becomes more challenging.

98 — Performance and Accuracy Vs Resolution

❑ Processing larger images requires larger networks which incur higher computational cost

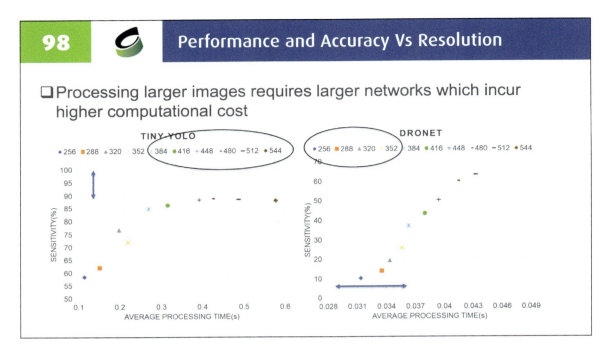

Input Image Size 960x540

Increasing the image size comes at a cost of reduced processing time while for small networks we have higher processing time but need to further increase the image size.

This is evident from the following graphs where we compare two networks in terms of image resolution Vs Performance.

Things get worse for smaller networks. Higher resolution is needed.

Performance is on CPU

99 — Designing Neural Networks for Edge Intelligence (I)

❑ **Deep Neural Networks**
  - used for decision making and data analytics
  - high computation and memory requirements

❑ **Current Trends**
  - reduce the bit-width precision of the network parameters and the processed data
  - use binary values
  - use compression techniques with quantization and pruning to reduce the network demand for memory and storage
    - [1]XNOR-Net
    - [2]Deep Compression

[1]M. Rastegari, V. Ordonez, J. Redmon, and A. Farhadi, "Xnor-net:Imagenet classification using binary convolutional neural networks,"
[2]Han, H. Mao, and W. J. Dally, "Deep compression: Compressing deep neural network with pruning, trained quantization and Huffman coding"

99 Designing Neural Networks for Edge Intelligence

- **Neural networks** are increasingly being used in many applications to provide decision making and data analytics (CNNs for computer vision tasks).
  - **Artificial Intelligence(AI)** is currently facilitated using machine learning algorithms such as **Deep Neural Networks(DNN)** which demands high computation and memory requirements.
  - Therefore, the computation was pushed from device to remote computing infrastructure.

Current trends for designing Neural network for the edge are trying to reduce both memory and computational requirement by reducing the bit-width precision of both the network parameters and processed data

Another techniques is to use binary values for the representation of the weights in order to avoid floating point operations which are high computational demanding

Lastly compression techniques with quantization and pruning of the network has show significant impact of both memory and storage

100 Designing Neural Networks for Edge Intelligence (II)

- ❑ **Optimize Neural Network architecture (DroNet Architecture)**
  - Achieved 35 FPS on a PC-based CPU platform
  - Trained with a custom database for vehicle detection
  - Processes 512 x 512 images
  - Make use of 3x3 filters and cheaper 1x1 convolutions
  - Progressively reduce the feature maps size by a factor of 2
  - Lower number of filters at early layers

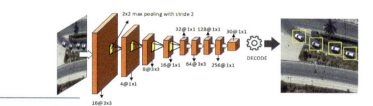

Christos Kyrkou, George Plastiras, Stylianos Venieris, Theocharis Theocharides, Christos-Savvas Bouganis, "DroNet: Efficient convolutional neural network detector for real-time UAV applications," 2018 Design, Automation & Test in Europe Conference & Exhibition (DATE), Dresden, Germany, pp. 967-972, March 2018.

Recap – the first approach was just an application-specific CNN with device (and network) specific optimizations.

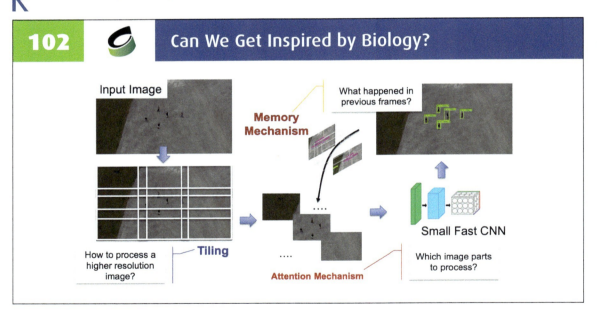

101 Designing Neural Networks for Edge Intelligence (III)

Reducing the image size effectively reduces the object resolution and detail
- Difficult to detect smaller objects such as pedestrians especially for small DNNs (DroNet/tinyYolo)

Resizing is NOT a good thing. You lose information, (and valuable time).

102 Can We Get Inspired by Biology?

An object detection algorithm for UAVs that:

- Can process higher resolution images without significant computational cost.
- Discard information and avoid unnecessary computations
- Avoid reducing the image accuracy and distorting the objects
- Make smaller objects detectable
- Feedback Mechanism

- The proposed algorithm is based on separating the input image into smaller regions capable of being fed to the CNN in order to avoid resizing the input image and maintain object resolution.
- Methodology that operates on higher resolution images.
- Detect smaller objects such as pedestrians with increased accuracy with no significant computational cost.

103 Tiling

- Uniformly distribute them across the input image to achieve complete coverage of the image while maintaining a constant overlap between the tiles.
  - Maintain object resolution
  - Does not distort image

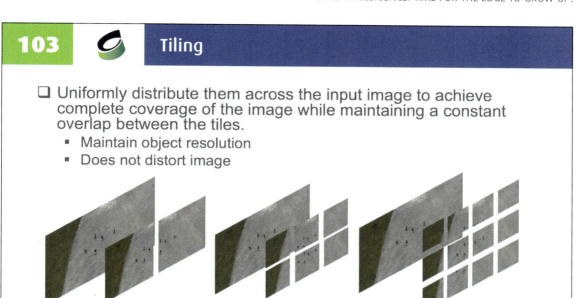

Essentially the idea is that the image is uniformly distributed in tiles, with a specific overlap to ensure that no information is lost. Then, find an intelligent method to choose which tiles will be evaluated for objects of interest.

104 Generation of Tiles

- Calculate the number of tiles that can be generated in the horizontal and vertical direction for a give input image

$$Ratio = \frac{Input_{size}}{CNN_{size}}$$

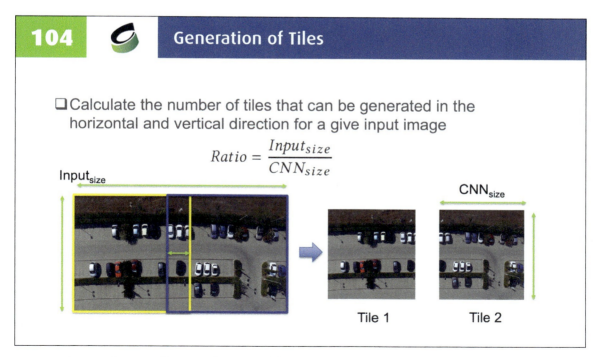

Ensure that overlap Is sufficient – determined by Johnson's criteria and also estimated size of object of interest. There is an advantage – drones have a barometer sensor which provides altitude – which is also the distance of a camera. Hence, object size can be estimated.

105 Memory Mechanism

- Keep track of detection metrics in each tile over time
    - Relative position of objects will not change significantly over a few successive frames
- A memory buffer is introduced that keeps track of various detection metrics within a tile
    - position of the bounding box with respect to the image
    - a detection counter for each bounding box
    - the latest tile that it was detected in
    - detection confidence
    - and the class type (vehicle, pedestrian etc.)
- Whenever a bounding box with high confidence is established it must first be categorized as new or belonging to an already
    - A stored bounding box with no overlapping detections after a certain number of frames is removed.

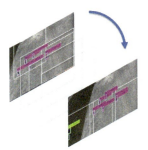

So, lets first develop some sort of history/memory mechanism which will enable us to turn our attention to the most appropriate tiles.

106 Memory Mechanism (2)

- The association between bounding boxes is done using the IoU and the class confidence.
    - High IoU implies high correlation with the previously stored box
    - Low IoU implies change in the scene and thus the importance of that tile should increase

- Keep the length of the memory buffer relatively small
    - .e.g look at the history of the previous 3 frames

We can use the Intersection over Union and the Class confidence. If there is a high IoU it implies that there's high correlation with the previous box. Low implies that the scene contains changes, and we need to reevaluate the tile.

107 Selective Tile Processing[1]

- Methodology that operates on higher resolution images
- Detect smaller objects such as pedestrians with increased accuracy with no significant computational cost
- Separating the input image into smaller regions capable of being fed to the CNN in order to avoid resizing the input image and maintain object resolution (Tiling)

[1] George Plastiras, Christos Kyrkou, Theocharis Theocharides, "Efficient ConvNet-based Object Detection for Unmanned Aerial Vehicles by Selective Tile Processing ", In International Conference on Distributed Smart Cameras (ICDSC), September 2018.

If we can determine a methodology to selectively process the tiles, we can save up a lot of time/computation. This was the first attempt.

108 Attention Mechanism

❑ Select which tiles to be process by the CNN
- All-Tiles – **TA**
- Single Tile (Round Robin) – **T1**
- Tiles with Previously Detected Objects – **TO**
- Tiles Selection with Memory – **TSM**

TO:
1. Process all frames
2. Select those with objects
3. When reset timer then search all again
   Derive a mechanism which will steer the algorithm towards selecting the high-probability tiles.

109 Tiles Selection with Memory – TSM

- There are four main criteria to asses the value of tile i
  - The number of objects detected in each tile – O_i
  - The cumulative intersection-over-union between current and previous bounding box locations – I_i
  - The number of times not selected for processing over time – S_i
  - The number of frames past since selected for processing – F_i
- Select top N tiles above a threshold for processing

$$V_i = \frac{O_i}{\max_j(O_j)} + \left(1 - \frac{I_i}{\max_j(I_j)}\right) + \frac{S_i}{\max_j(S_j)} + \frac{F_i}{\max_j(F_j)}$$

$$\forall j \in [0, \ldots, (N_T - 1)]$$

How recently we processed the tile vs how frequent we processed the tile!

110 Evaluation and Experimental Results

- Analysis on an embedded CPU platform (Odroid XU4)
  - Target Application is pedestrian detection from UAVs
  - UCF Aerial Action Data Set (2011)
- Different configurations with different base CNN and tile selection strategies

Tiny-YoloV2	DroNet^{TA}	DroNet^{TO}
DroNet	DroNet^{T1}	DroNet^{TSM}

All-Tiles – **TA** ---- Single Tile (Round Robin) – **T1**
Tiles with Previously Detected Objects – **TO** ---- Tiles Selection with Memory – **TSM**

$$SEN = \frac{T_{pos}}{T_{pos} + F_{neg}} \qquad ATP = \frac{1}{N_{test\_samples}} \times \sum_{i=1}^{N_{test\_samples}} t_i$$

111 Tile Selection Metrics

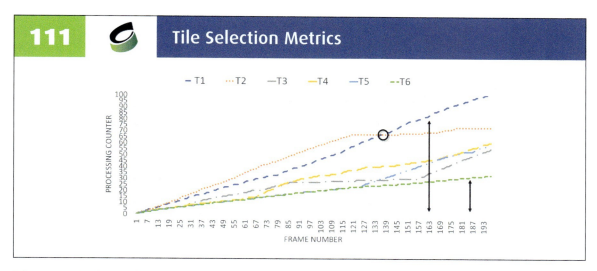

The ratio of selected tiles is relatively constant
  Maintain fairness between tiles
Tiles that are selected more often correspond to those with more pedestrians
  Reset counters every 10 frames
  Process tiles based on number of objects
   - Sensitivity is above 88%
   - 2× improvement in processing time than processing all tiles
  Process tiles based on selection criteria

- Number of selected tiles is below 35% on average
- 2-3× improvement compared to processing all tiles
- Maintain comparable performance to single CNN with resizing
- Managed to improve sensitivity between 20-70%
- Managed to find balance between the extreme cases

112 Accuracy Vs CNN Input Size

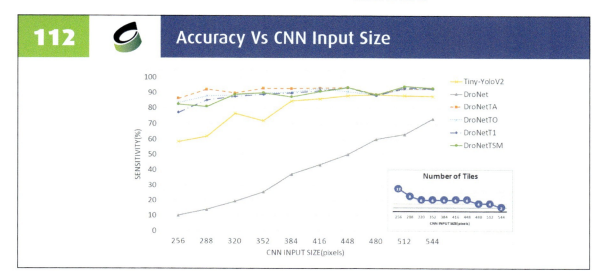

All approaches improve sensitivity Even at the worst case
Baseline Approaches
 - Sensitivity ranges from 88% to 58% for larger images
 - In order to obtain higher sensitivity higher resolution images must be used.
Processing all Tiles
 - Higher sensitivity between 86-93%
Single Tile processing
 - For small number of tiles (2-6) sensitivity is maintained above 75%

113 Processing Time Vs CNN Input Size

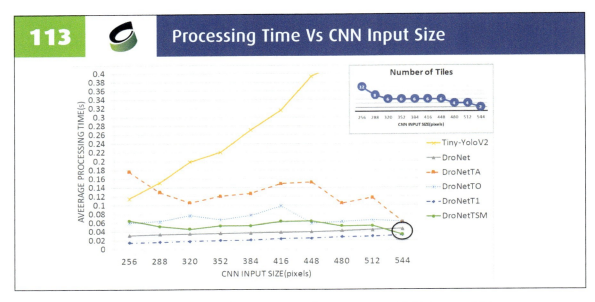

Can be as fast as single image Dronet. No resizing at the end.

Faster when not resizing with only two tiles and swapping.

Even processing all tiles the approach is faster than tinyYOLO with higher sensitivity.

Processing all tiles
- Simultaneously processing time increased 5× per frame

114 Performance on Odroid XU4

❑ Experimental analysis on Odroid XU4 embedded platform on a DJI Matrice 100 UAV for **pedestrian detection**.

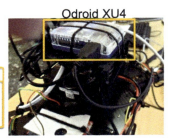

Approach	Average Processing Time(sec)	Sensitivity(%)
Tiny-YoloV2	1.646	63.512
DroNet	0.207	19.863
DroNet^{TA}	0.962	92.379
DroNet^{T1}	0.107	80.101
DroNet^{TO}	0.482	88.145
DroNet^{TSM}	0.394	91.109

George Plastiras, Christos Kyrkou, and Theocharis Theocharides. 2018. Efficient ConvNet-based Object Detection for Unmanned Aerial Vehicles by Selective Tile Processing. In *Proceedings of the 12th International Conference on Distributed Smart Cameras* (ICDSC '18). ACM, New York, NY, USA, Article 3, 6 pages. DOI: https://doi.org/10.1145/3243394.3243692

Input is 256 with 12 tiles.

Detection Results

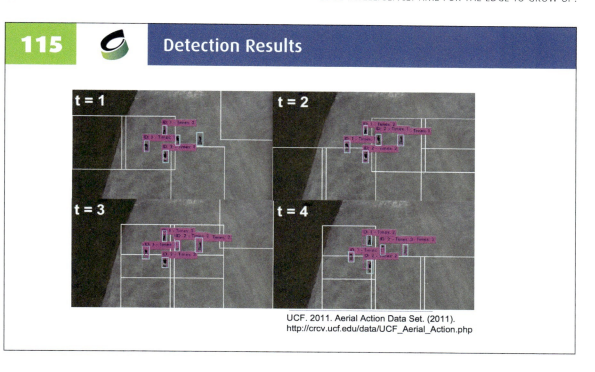

UCF. 2011. Aerial Action Data Set. (2011).
http://crcv.ucf.edu/data/UCF_Aerial_Action.php

Selective Tile Processing

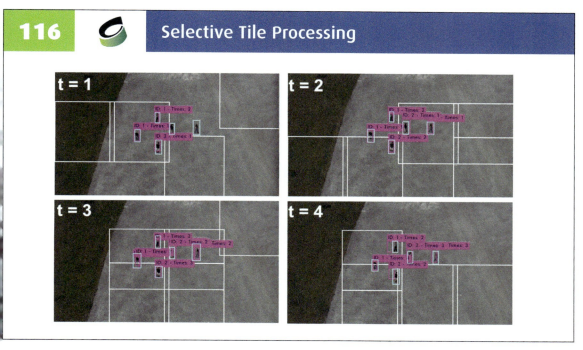

The slide explains visually how tiling works and also what are the current drawback.

117 Impact of Tiling, Memory & Attention

- Through tiling the sensitivity is improved.
- By intelligently selecting the tile to process performance is not impacted as heavily
- Any of the techniques can be utilized for the given application with any underlying CNN
- Can further tune parameters and selection criteria to achieve even higher performance

- Tiles are static → Dynamically adjusting the number, the size, and positioning of the tiles based on the recorded activity
- A lot of room for improvement

Overall tiling improves performance. Highlighted different approaches. The best one is dependent on each application.

The CNN can also change depending on application demands.
A lot of room for improvement!

118 EdgeNet Framework Overview

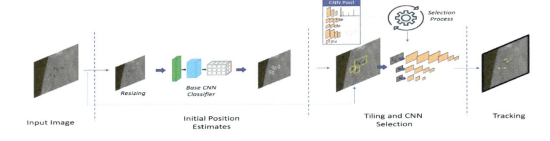

Hence, we present EdgeNet. A framework that consist of multiple stages with a goal to optimize both performance and the accuracy of a detector

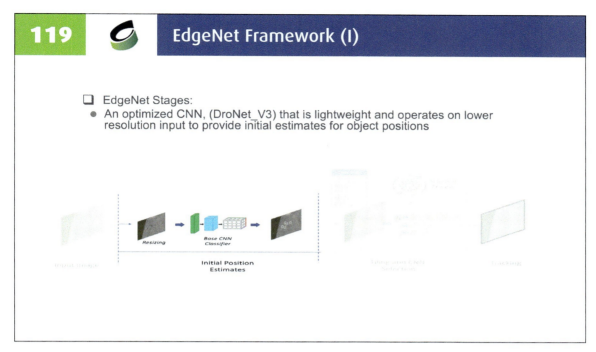

Shortcut connections
- Feature map up sampling and concatenation

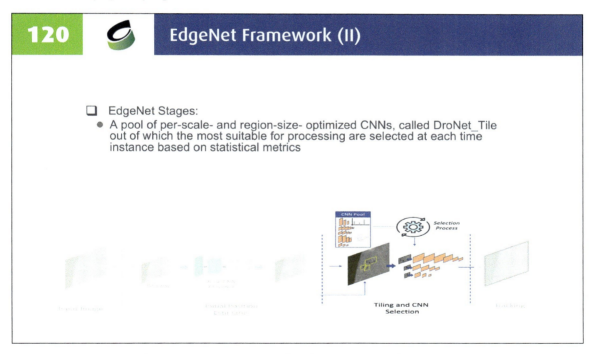

What if we use a pool of CNNs, that are matched to the appropriate tile?

EdgeNet Framework (III)

- EdgeNet Stages:
  - An optical-flow tracker to compensate for the increasing demands of the previous stages and speed-up of the whole process

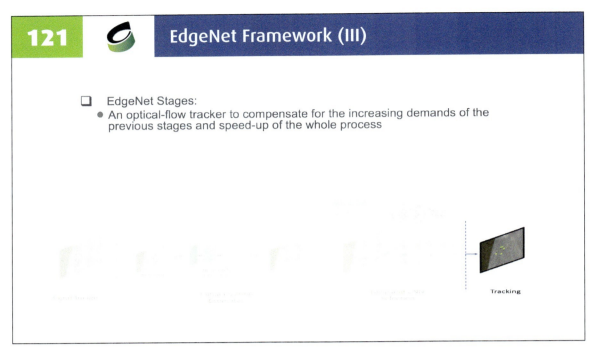

Take advantage of temporal aspects through tracking

Initial Position Estimation

- Responsible for producing the initial positions of objects
- Use an efficient CNN designed for Edge applications (DroNetV3)
- Works with the traditional way of resizing the input image – Select larger input

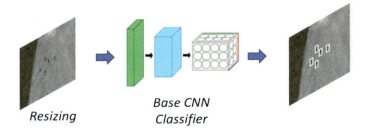

We first find our objects – this happens only once though during each iteration.

123 Tiling and CNN Selection (I)

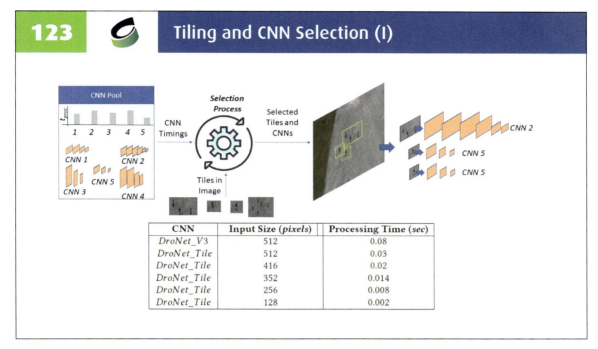

First we run a profiling of all the cnns in the cnn pool.

In our case we have the DroNet_V3 and the original DroNet, which we call DroNet_Tile with different input sizes.

124 Tiling and CNN Selection (II)

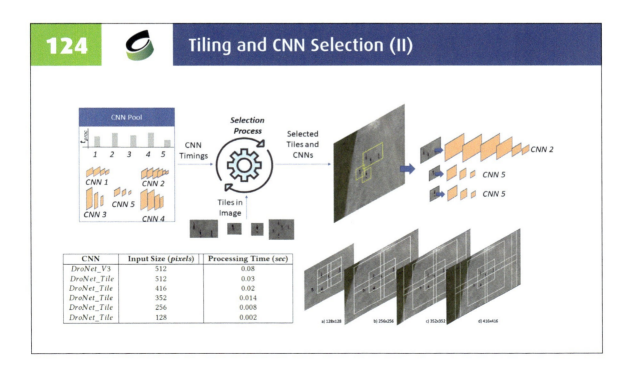

124 Tiling and CNN Selection (II)

For each detected box proposed by the first stage a number of tiles are generated by positioning the object at each of the four tile corners, as shown.

In addition, tiles with different sizes are also generated, in our case we used a total of 5 sizes: 512, 416, 352, 256, 128 matching the different sizes in the CNN pool, as shown in the table. A total of 20 tiles for each object are proposed, where each tile is evaluated by the selection process based on the objects that it covers and its associated processing time. Thus, for each of the 20 tiles per object proposed we calculate an Effective Processing Time (EPT), which is the number of objects that are covered divided by its corresponding processing time (Table 1).

From the proposed tiles per object we select the one with the minimum EPT.

Finally, we combine all the extracted tiles for all objects, and discards the redundant ones (i.e., those that cover the same or fewer objects) and retain only the one with the minimum EPT.

125 Optical flow (Lucas-Kanade[1] tracker)

- Works on the principle that the motion of objects in two consecutive images is approximately constant relative to the given object
- Tracker is used to:
    - track the objects of the framework along with stage 1 and 2 and compare and verify the position of the detected object using both tracking and detection algorithms
    - reduce the processing time of the framework using only the tracker, before detecting the whole image again.
- Extract centered points of each detected box
- Use previous and current frame to calculate the optical flow of the points and return the estimated new position of each object

[1] Bruce D. Lucas and Takeo Kanade. 1981. An Iterative Image Registration Technique with an Application to Stereo Vision. In Proceedings of the 7th International Joint Conference on Artificial Intelligence - Volume 2 (IJCAI'81). Morgan Kaufmann Publishers Inc., San Francisco, CA, USA, 674–679

As a tracker, we can use any good tracker with minimal code footprint/complexity which is why we resented to Lucas-Kanade.

126 Experimental Setup

- Images were collected using manually annotated video footage from a UAV and the UCF Aerial Action Data Set
- Training set consist of 1500 images with a total of 60000 pedestrians
- We trained each CNN on a Titan Xp GPU with 256 batch size, 0.001 learning rate and 0.005 decay.
- We used Darknet, a C- and CUDA- based Neural Network framework, to train, test and evaluate each CNN on different platforms
- Evaluated on 198 sequential images containing 988 pedestrians in total
- We used three different computational platforms a low-end Laptop CPU, and then ported on two embedded platforms an Odroid[1] device and a Raspberry Pi3[2]

[1]Samsung Exynos-5422 CortexT M-A15 2Ghz and CortexT M-A7 Octa-core CPUs with Mali-T628 MP6 GPU
[2]RPI3: Quad Core 1.2GHz Broadcom 64bit CPU

How we evaluated the approach!

127 Time-slot Selection

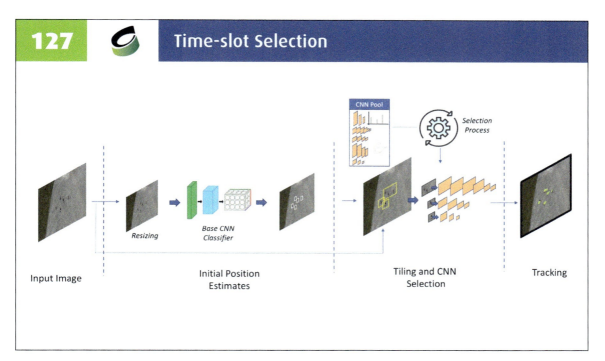

This figure shows the timing of the detector.

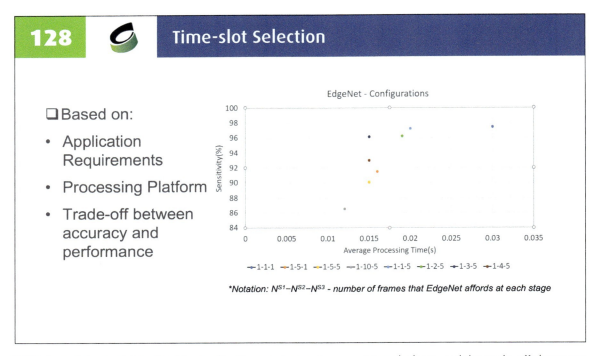

The time slots are determined by application requirements, processing platform, and the trade-offs between accuracy and performance.

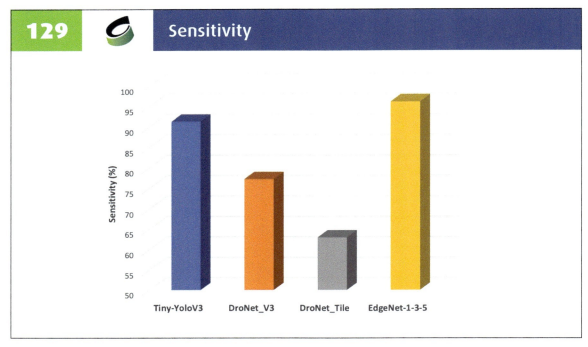

Results for Sensititvity

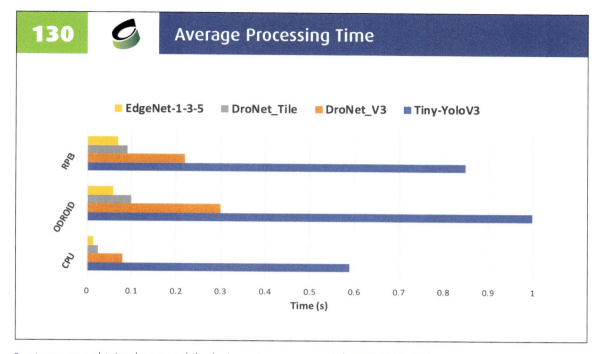

Latency was obtained on a mobile device using: Samsung Exynos-5422 CortexT M-A15 2Ghz and CortexT M-A7 Octa-core CPUs with Mali-T628 MP6 GPU Quad Core 1.2GHz Broadcom 64bit CPU

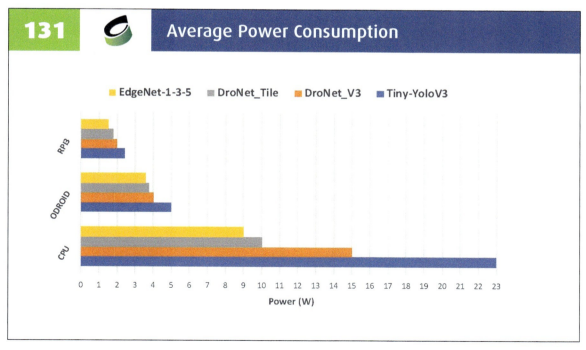

Obviously the power results are important when dealing with the edge devices.

132 DroNet in ACTION (I)

Examples of the detector.

133 DroNet in ACTION (II)

DroNet on CPU

DroNet on Android

Version 1 and V3.

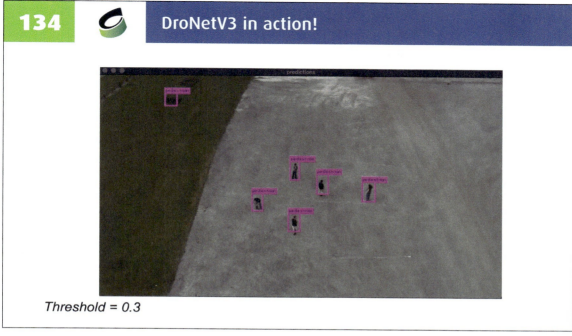

Dronet V3.

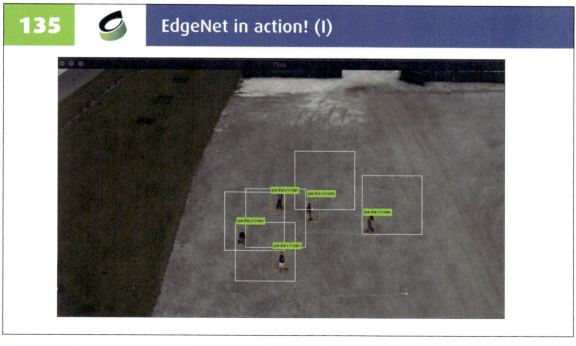

Edgnet has better accuracy while less energy and more performance!

136 — EdgeNet in action! (II)

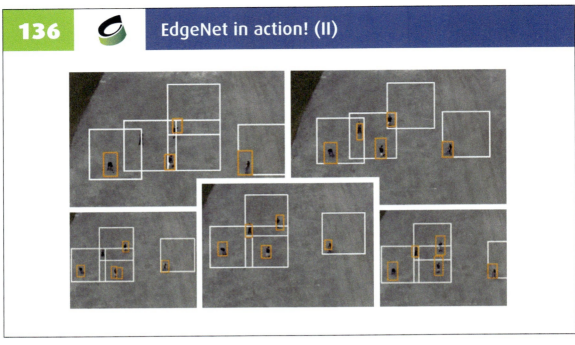

Here are the bounding boxes.

137 — Observations

- **Application Needs Assessment:**
    - Sometimes simple solutions maybe sufficient!
        - **Don't believe the hype! If the data & sampling is small, go small!**
    - Holistic Parameter/Space Exploration
- **If we do need Deep Learning…**
    - Existing CNN models limited? → Exploration of CNN models
    - Formalizing performance metrics and application requirements
    - DroNet an example of such methodology
        - Provides real-time performance
        - Comparable accuracy to other networks on edge
- **Bio-Inspired Optimizations**
    - Memory & Attention Mechanisms related to human saliency
    - Sometimes we don't need all the information

PART 4

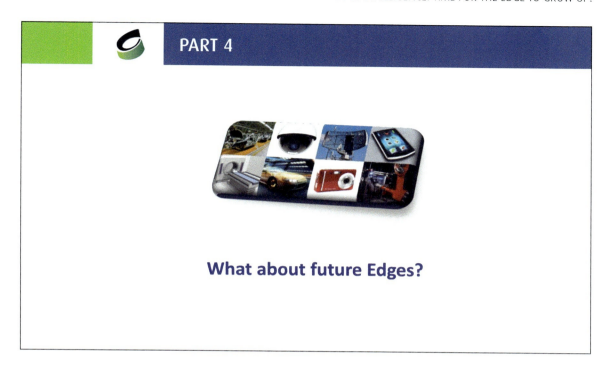

What about future Edges?

138 — Post Von-Neumann Computing

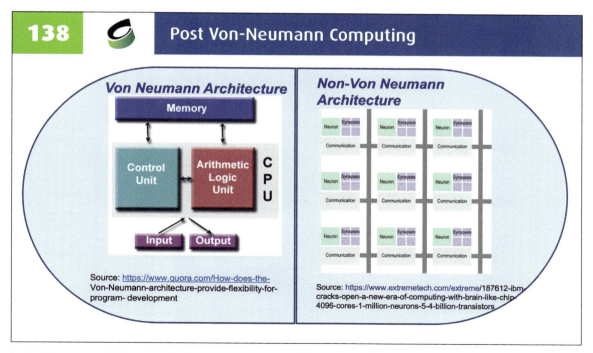

Source: https://www.quora.com/How-does-the-Von-Neumann-architecture-provide-flexibility-for-program-development

Source: https://www.extremetech.com/extreme/187612-ibm-cracks-open-a-new-era-of-computing-with-brain-like-chip-4096-cores-1-million-neurons-5-4-billion-transistors

Von Neumann Architectures lack the appropriate capabilities to further boost convolutional and other neural networks. Especially for the edge, they are not as good as other architectures.

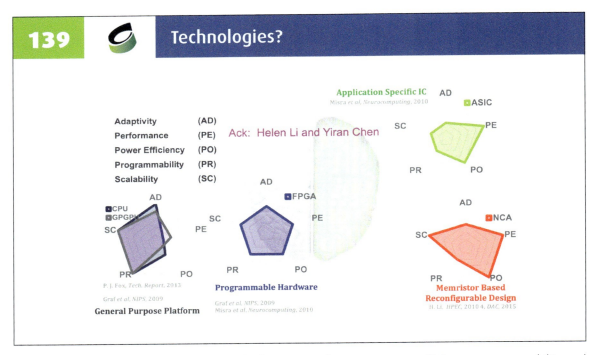

A comparison of technologies in terms of adaptivity, performance, power efficiency, programmability and scalability.

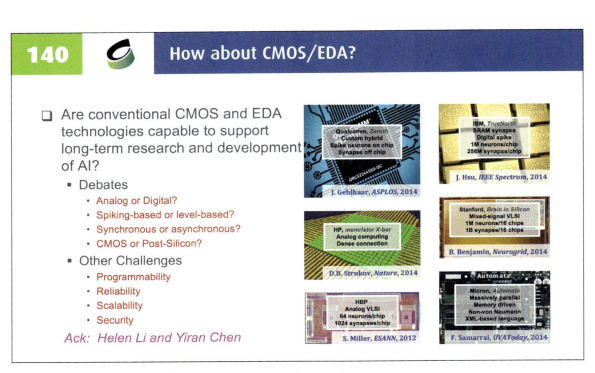

But, what about the current CMOS technologies, and what about EDA?

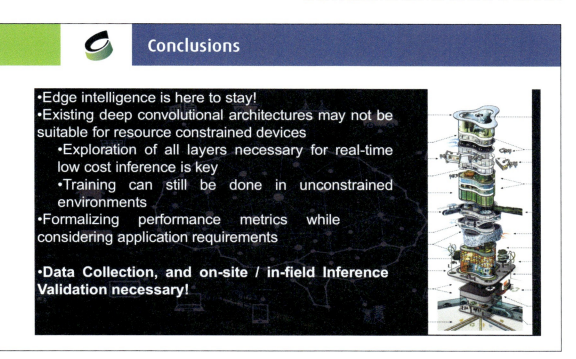

To conclude: Edge Intelligence is here to stay and we need to embrace it and focus on building energy efficient, cost-effective, robust and real-time AI algorithms

The whole design cycle followed in traditional circuits and processor architectures is more or less the same, but:

Need IN FIELD Evaluation. There are several parameters that change when going from simulation to evaluation.

CHAPTER 05

Putting the Humans in the Middle of the CPS Design Process

Anna-Maria Velentza

University of Macedonia, Greece

#	Title	Page
1.	Market Size Expand	178
2.	Human in the Loop	179
3.	Design efficient Cyber Physical Systems The road so far...	179
4.	Human-Machine Abilities	180
5.	Human-Machine Collaboration	181
6.	The Automotive Car Example	182
7.	The automotive car example	182
8.	Source of Theories for Modeling Human Behavior	183
9.	Emotions-Facial Expressions	184
10.	Charles Robert Darwin (I)	184
11.	Charles Robert Darwin (II)	185
12.	Paul Ekman	186
13.	Paul Ekman's Experiment	186
14.	Paul Ekman (II)	187
15.	Theories of Emotions and Computational Systems	188
16.	From Psychological Theories to... even FPGAs	188
17.	Body Language	189
18.	5 Types of Nonverbal Behavior (I)	189
19.	5 Types of Nonverbal Behavior (II)	190
20.	5 Types of Nonverbal Behavior (III)	190
21.	5 Types of Nonverbal Behavior (IV)	191
22.	5 Types of Nonverbal Behavior (V)	191
23.	Principles of Emotional Expressions According to Darwin	192
24.	Principle of Serviceable Habits	192
25.	Principle of Antithesis	193
26.	Principle of Nervous Discharge	193
27.	Progress in Emotional Computing	194
28.	HCI Norms (I)	195
29.	HCI Norms (II)	195
30.	How to use theories for Modeling Human Behavior	196
31.	Human Behavior Modeling- Methods	197
32.	Memory Process (I)	198
33.	Memory Process (II)	198
34.	Types of Memory	199
35.	Human Behavior Modeling- Physical Reactions	200
36.	Human Behavior Modeling-Scenarios	200
37.	Related Work on Lighting Conditions	201
38.	Using Memory theories for HCI	202
39.	Task-Procedure (I)	202
40.	Task-Procedure (II)	203
41.	Task Example	204
42.	Results	204
43.	Use of the Results	205
44.	Discussions	206
45.	Bibliography	206

The Market of IoT and Cyber Physical Systems (CPS) is growing and comprises of several stakeholders, from the conventional hardware and software designers to healthcare and education industries. One crucial aspect which will allow CPS to continue growing in to integrate more humans in the design loop; the design of CPS should consider how humans interact with the physical world, for example how they communicate or move. Humans, and machines (hardware and software) must interact and understand each other in order to work together in an effective and robust way. A trustworthy source of theories about human's emotions, behaviours and cognitive situation that can be utilized so as to enhance human-machine interaction stems from the field of Psychology. In the current chapter we are discussing examples of emotional theories, evidences on how people prefer to interact with intelligent machines and valid and reliable ways to collect data in order to model human behaviour in order to design high efficient human-centric CPS.

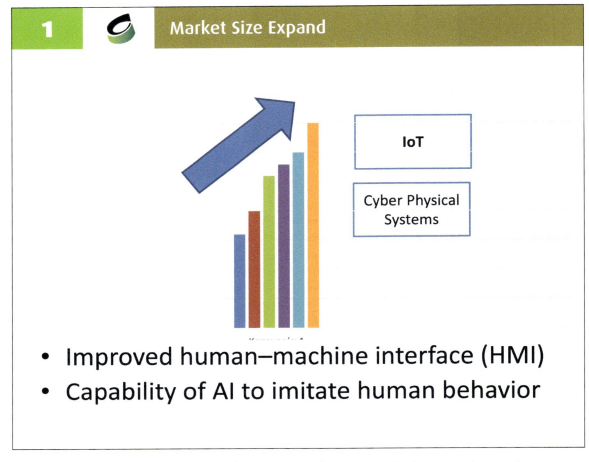

The Market of IoT and Cyber Physical Systems is a growing market that comprises several stakeholders, from the conventional hardware and software designers to even healthcare and education industries. This also works in the opposite direction; increasing applicability of cyber physical systems in multiple industries including healthcare, energy, automation, building design, agriculture, manufacturing, and transportation is likely to assist in the expansion of the global cyber physical system market.

The cyber physical systems market is geographically spread in the following regions: -United States, Europe, China, Japan, Southeast Asia, and India. During 2019, Cyber Physical System industry is much fragmented, manufacturers are mostly in the North America and Europe. Among them, Europe Production value accounted for less than 29.75% of the total value of global Cyber Physical System [1].

The global market for IoT discrete parts manufacturing was valued at $643.9 million in 2016 which has grown at $862.6 million in only one year. From 2017 it is estimated to grow to $2.8 billion by 2022 with a compound annual growth rate (CAGR) of 26.7% for the period of 2017-2022. As for the global market for IoT sensors, it is projected to grow from $14.1 billion in 2018 to $48.0 billion by 2023, at a compound annual growth rate (CAGR) of 27.8% from 2018 to 2023.

2 Human in the Loop

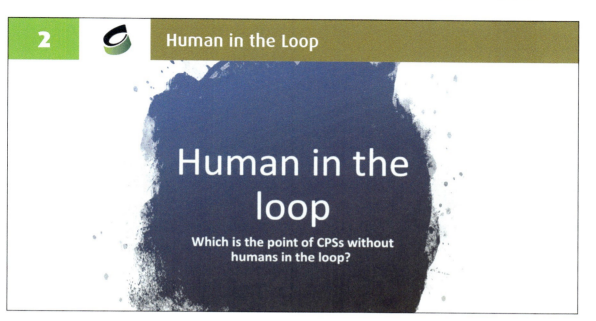

Human in the loop

Which is the point of CPSs without humans in the loop?

It is crucial for Cyber Physical Systems (CPS) to continue their annual growth by keeping humans in the loop. The design of CPS will/should take into account how humans interact with the physical world, for example how they communicate or move.

Humans, and machines (hardware and software) must interact and understand each other in order to work together in an effective and robust way. Even in full autonomous scenarios (e.g. fully autonomous cars, robots at factories of the future, etc.), systems are required to explain their behavior to humans, provide them with feedback and allow them to override system actions. Apart from that, all those autonomous scenarios will become plausible only after we have fully understand and model human behavior and human interaction with the physical world. A trust-based relationship must be built between humans and machines so they can work together and achieve effective human-machine integration [3].

3 Design efficient Cyber Physical Systems – The road so far...

- **Research on CPS mainly focus on:**
  - Hardware
  - Software
- **Research on human-centered CPS focus on:**
  - Not physically dangerous
  - Protection from cyber attacks (Security)
  - Models of Human Behavior
  - Feedback
  - Control
  - Autonomic Computing

3 Design efficient Cyber Physical Systems The road so far...

The research on CPS mainly focus on Hardware and Software, including devices for a remote dashboard, devices for control, servers, a routing or bridge, sensor devices as well as energy modules, power management modules, RF modules, and sensing modules. On the other hand, most of the research in CPS with humans in the loop focuses on the need for a comprehensive understanding of the complete spectrum of the types of human-in-the-loop controls, on the need for extensions to system identification or other techniques to derive models of human behaviors, and also, on determining how to incorporate human behavior models into the formal methodology of feedback control [4]. Those CPS' challenges incorporate the essential characteristics of operational mechanisms which connect the physical network, cyberspace, mental space, and social networks [5]. Moreover, it is important to take into consideration the aspect of security, -cyber and physical- by creating systems that will not be physically dangerous and ready to protect both the humans and the whole system from cyber attacks.

Finally, Autonomic Computing is another field of research essential to design systems with self-managing capabilities. These systems apply control theory principles to make use of control loops.

4 Human-Machine Abilities

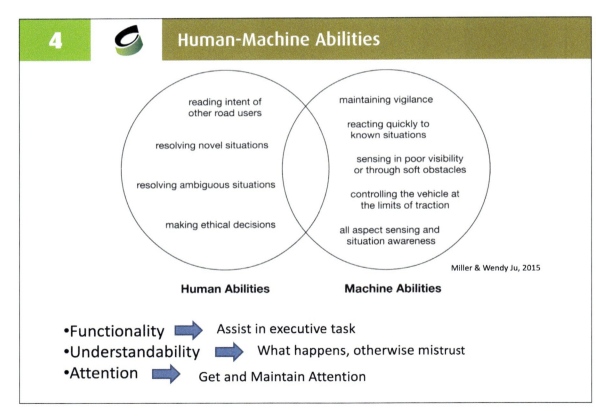

Humans and Machines have different abilities although there are partially overlapping areas as you can seen in the road vehicle systems example [6]. Humans are better on reading the intent of other road users, resolving novel and ambiguous situations and making ethical decisions. Machines are better, compared to humans, in maintain vigilance, reacting quickly to known situations, sensing in poor visibility or through soft obstacles, controlling the vehicle at the limits of traction and generally in all aspects of

4 Human-Machine Abilities

sensing and situation awareness.

In order humans to feel comfortable during human-machine interaction and use all the benefits of the machine abilities there are 3 specific prerequisites that must be met.

Complement functionality. Humans are integrated in an intimate collaboration with the system to assist in the execution of certain tasks in a correct and complete manner.

Achieve understandability. A human must understand how the CPS operates and what is happening at the current moment or even the moment before, otherwise, the human may mistrust the system. It is necessary to make use of mechanisms that provide relevant feedback and feedforward to the human. In case that the humans do not understand the machines and they do not trust them, they are not taking into consideration the machine's warnings or suggestions.

Manage user attention. In order to have a functional collaboration between humans and machines, machines needs to get or maintain the human's attention to *the CPS*. The human can be distracted or focused on another task at the moment the *CPS* could/should require her/his participation. [3]

5 Human-Machine Collaboration

- The automotive car example

- Navigation
- Distance from passing vehicle

- Fatigue
- Attention
- Alcohol
- Emotional State

- Road Conditions
- Weather Conditions

In order to explain better the Human-Machine collaboration, we can use the automotive car example. Driving is an attention-demanding task and as the complexity of utilizing the new features of the vehicles increases, there is an ever more pressing need for safe interfaces that support drivers while they are multi-tasking. As a result, driving assistant systems are becoming attractive in the field of intelligent robotics/ systems. In the automotive domain, humans meet a variety of situations while driving cars and intelligent systems help considerably humans to understand how surrounding situations change.

There are three main aspects in such a CPS; the vehicle (i.e. Navigation, distance form passing vehicle), the environment (i.e. weather conditions, road conditions such as slippery of the road) and the human-driver (i.e. fatigue, attention level, alcohol consumption, emotional state).

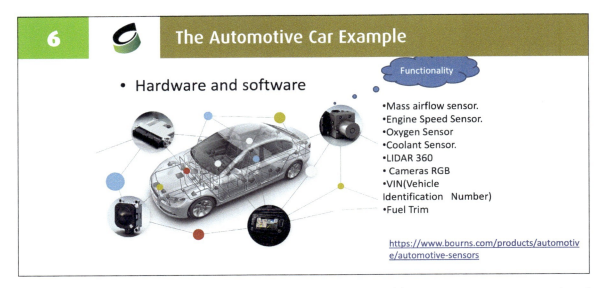

There are many hardware and software solutions for monitoring the vehicles' and the roads' conditions. From specific sensors that are embedded, the vehicle receives data and will inform the driver for external conditions (e.g. road slippage, distance of vehicles passing, lane identification and unwanted changes), appropriately and when needed. There are plenty of sensors such as mass airflow sensors, engine speed sensors, oxygen sensors, coolant sensors and many others that provide information about the vehicles' condition. Information such as VIN (Vehicle Identification Number), engine speed, vehicle speed, accelerator pedal position in%, MAF Rate and Fuel Trim that together estimate fuel consumption can be extracted directly through OBD II and help record useful system variables that will help extract driver features.

Roadway status will be fused by data from a number of different sources e.g. RGB cameras, beacons, accelerometers and encoders on the wheels of the vehicle, implementing and evolving existing methods [7]. All those sensors and the information retrieved from them serves the **Complement functionality prerequisite** .

7 The Automotive Car Example

Humans can be deeply influenced by affection during social interaction. Specifically, emotional cues from others can be a powerful way to persuade people to modify their behaviors. Volkswagen and MIT utilized this approach in developing humanoid robots acting as driving assistants. For example, Foen and Williams explored how a social robot called AIDA (Affective Intelligent Driving Agent) can better persuade drivers to adhere to road safety guidelines as compared to existing technologies, and AIDA's persuasiveness has been proved much higher when compared to that triggered by a human co-passenger. From the viewpoint of interaction performance, proper situation awareness by a robotic assistant in a vehicle and the relevant determination of corresponding reactions are crucial prerequisites for long term interaction between a human driver and a vehicle's robotic system. Yang focus on preserving human-robot interaction for driving situations, considering how many types of different cognitive situations occur and how affective interaction can be triggered by the robotic driving assistant. By gaining human trust we serve the aspect of **Understandability**. A trust-based relationship must be built between humans and machines so they can work together and achieve effective human-machine integration.

8 Source of Theories for Modeling Human Behavior

- **Theories comes from the field of Psychology**

- Bhatia, Sudeep, Russell Richie, and Wanling Zou. "Distributed semantic representations for modeling human judgment." *Current opinion in behavioral sciences* 29 (2019): 31-36.
- Hasse, Cathrine, and Dorte Marie Søndergaard, eds. Designing robots, designing humans. Routledge, 2019.
- R. Ogawa. "How Humans Develop Trust in Communication Robots: A Phased Model Based on Interpersonal Trust." 2019 14th ACM/IEEE International Conference on Human-Robot Interaction (HRI). IEEE. 2019.

However, where can we find theories in order to model human behaviour? A trustworthy source of theories about human's emotions, behaviours and cognitive situation stems from the field of Psychology. Many theories that are applied in the engineering field such as computer vision are based on psychological and neuroscience findings about how the human brain, vision system and behaviour works.

9 Emotions-Facial Expressions

Inside Out 2015 [11]

Based on that, we can have a brief look on emotional theories and how we are using them today for modeling human behavior.

10 Charles Robert Darwin (I)

- 1809-1882
- Theory of Evolution
  - Natural Selection
    - Not only for physical characteristics but also for mental (cognitive mechanisms, emotions…..)
  - Emotions are heritage of our evolution
  - Although some emotions and expressions are not used in modern life

10 Charles Robert Darwin (I)

Charles Robert Darwin (1809-1882) is well known for the Theory of Evolution by natural selection. Natural selection is the process describing how the species/organisms' characteristics were able to adapt to the challenges of the environment, survive the dangers and give birth to offspring's. Such characteristics are for example the tall neck of the giraffes that makes it possible for them to eat from the high trees or the duck feet that make it possible for them to walk and to swim by using them. The theory is sometimes described as "survival of the fittest" which is not referred to an organism's strength or athletic ability, but rather to the ability to survive and reproduce. The natural selection does not cover only the physical characteristics but also the mental ones such as cognitive mechanisms and emotions.

In 1872 Darwin postulated in his original work that facial expressions were the results of improvements and that the study of expressions is difficult owing to the movements (of facial muscles) being often extremely slight and of fleeting nature. Emotions come together with our evolution despite the fact that some emotions and expressions are not used in modern life.

11 Charles Robert Darwin (II)

There are some facial expressions that indicate the emotional state of the species. The animals by moving their muscles can communicate their emotions and intentions. Our ability to understand their emotional state makes it possible for us to protect ourselves from their attack (in the case of the upper pictures) or to protect the animals in the pictures at the bottom. We can consider puppies as cute but the reason is because they need our protection (they are unable to protect theirselves) and so their physical characteristics trigger the instincts of protection.

12 Paul Ekman

- 1934-...
- 6 Basic Emotions
  - Happiness
  - Sadness
  - Anger
  - Disgust
  - Fear
  - Surprise
- "Lie to me"
- Ekman's Facial Action Coding System (FACS)

Paul Ekman (1934-...) is an American psychologist and professor emeritus at the University of California, San Francisco. He has created an "atlas of emotions" with more than ten thousand facial expressions and his research proved that some emotions and facial expressions are universal. Based on his original research he ended up with 6 universal, cross-cultural emotions: Happiness, Sadness, Anger, Disgust, Fear, Surprise.

By studying emotions and facial expressions, Ekman adopted the **Facial Action Coding System (FACS)**, a taxonomy system for human facial movements, originally developed by a Swedish anatomist named Carl-Herman Hjortsjö. This system is based on the movements of each muscle in the human face, even to the tiny and short movements that happen unconsciously when someone has an expression.

His work is frequently referred to in the TV series *Lie to Me* (https://en.wikipedia.org/wiki/Lie_to_Me).

13 Paul Ekman's Experiment

13 Paul Ekman's Experiment

But what does really mean that emotions and facial expressions are universal?

Dr. Paul Ekman with his colleagues went to Papua New Guinea and met Fore who lived in an isolated, preliterate culture using stone implements which had never seen any outsiders before. In order to test his theory about the universal cross-culture facial expressions, he asked the Fore people to replicate some facial expressions. Since the Fore people did not have specific words to use about emotions, Ekman used some short stories and asked them how they feel and look like when (1) Friends had come. (2) His child had just died. (3) He was about to fight. (4) He stepped on a smelly dead pig as you can see in the pictures accordingly. He also gave them to mach photos of western people with those stories. He replicate the same methodology with western people. Results proved the universal emotional theory. Fore people replicated similar facial expressions to those of western people and both of them were able to recognize the emotional state of the others.

This collection that can be also seen in the photos above, was first published in the 1971 *Nebraska Symposium on Motivation* where Paul Ekman presented the Universal and Cultural Differences in Facial Expressions of Emotion.

14 Paul Ekman (II)

6 Basic Emotions
- Happiness
- Sadness
- Anger
- Disgust
- Fear
- Surprise

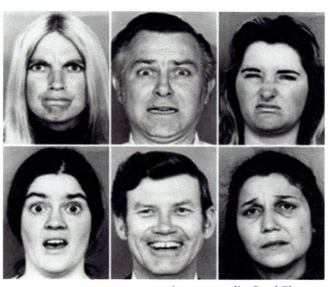

Image credit: Paul Ekman

Over here you can see and recognize the facial expressions from the 6 basic emotions replicated by western people.

15. Theories of Emotions and Computational Systems

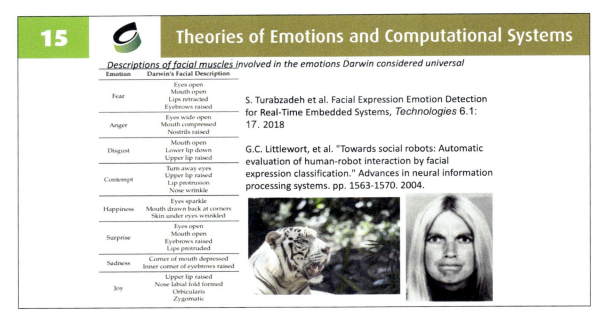

Going from the theory to application, Theories of Emotions are used in Computational Systems. In "Facial Expression Emotion Detection for Real-Time Embedded Systems" the authors used a description of the facial muscles involved in the emotions Darwin considered universal for automatic real time emotional detection. Here is an example that has also been used by Ekman : the emotion of anger, given Darwin's muscle description with eyes wide open, mouth compressed and nostrils raised.

16. From Psychological Theories to... even FPGAs

- **Automatic emotional state detection and analysis**

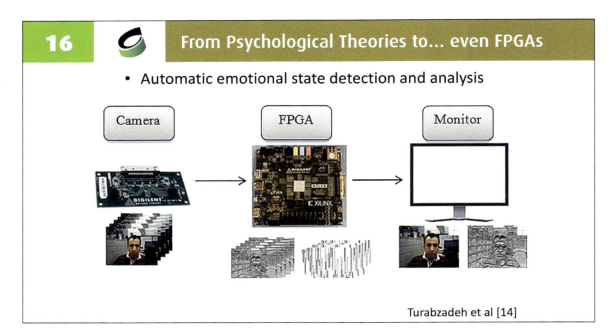

Turabzadeh et al [14]

By utilizing the theory of how to recognize emotions and facial expressions, we can design software and hardware solutions for real-time emotion detection with the aid of computer vision. [14]

17 Body Language

Humans do not communicate their feelings and cognitive situations only through their facial expressions but also through their whole body.

18 5 Types of Nonverbal Behavior (I)

- **Insignia**
  - Non verbal gestures that translates directly into words

There are five categories of nonverbal behavior that help people communicate their feelings. All those behaviors can vary according to the culture of the people that are using them.

First of all, there are the *Insignia* which are non verbal gestures that translates directly into words.

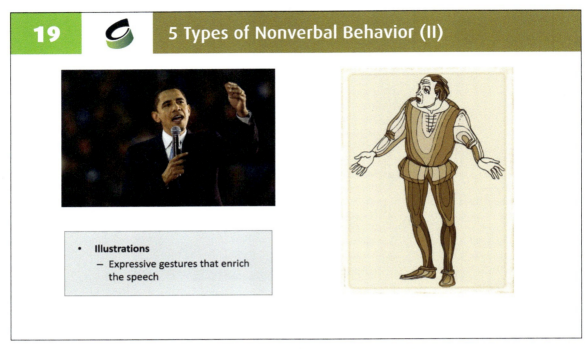

Secondly there are the *Illustrations* which are expressive gestures that enrich the speech.

Furthermore, people use *Regulators*, gestures that facilitate the management of situations such as the way a policeman manages the traffic.

21   5 Types of Nonverbal Behavior (IV)

- **Self-Adjustments**
  - Seemingly pointless nerve behaviors

Moreover, people can communicate their feelings with *Self- adjustments*, seemingly pointless nerve behaviors. People are using those in order to comfort theirselves in awkward or stressful situations. They are used as "reset" moves that help people relax.

22   5 Types of Nonverbal Behavior (V)

- **Non verbal emotional expressions**
  - Criteria
    - Short (1-10 milliseconds)
    - Spontaneous, Uncontrollable
    - There are corresponding expressions in other species of the animal kingdom

Finally, there are the non-verbal human emotional expressions which are short, in the range from 1 to 10 milliseconds, spontaneous and uncontrollable, like shy blushing while there are corresponding expressions in other species.

23 Principles of Emotional Expressions According to Darwin

- Principle of serviceable habits
- Principle of antithesis
- Principle of nervous discharge

The non verbal emotions according to Darwin are separated into three categories; 1) Principle of serviceable habits, 2) Principle of antithesis and 3) Principle of nervous discharge.

24 Principle of Serviceable Habits

Expressive behaviors that had a useful functional role were selected by the physical selection process and retained although they did not maintain appropriate functionality

- Visible to muscles, less subject to voluntary control

As we previously discussed about the nature of the emotional expressions, the *Principle of serviceable habits* is based on the theory that expressive behaviors that had a useful functional role and were selected by the physical selection process were retained although they did not maintain appropriate functionality. For example, nowadays we have more efficient ways to express our anger than showing our teeth but those expressions are not under voluntary control.

25 Principle of Antithesis

> When emotional situations arise in exactly the opposite way, they tend to be accompanied by expressive behaviors that are just the opposite, even if they do not have any real functionality.

Proud

Uncertainty

http://bodylanguageproject.com/

For example, pride and uncertainty are opposite emotions that serve to indicate success and failure respectively and by expressing them both the observer and the expresser benefit, based on the *Principle of Antithesis*.

Interestingly, the nonverbal displays are adaptive and confer certain degrees of fitness to both the displayer and the viewer. A common expression of pride is by making the body appear as large as possible, putting the shoulders back with an expanded upright posture, often by placing the hands on the hips. On the other hand, the feeling of uncertainty makes the body take on a smaller form by slumping the shoulders and lower the head. This is important because it tells us how the behaviour is favoured over time helping it to persist in the overall repertoire of human interactions.

26 Principle of Nervous Discharge

> Excessive sensory stimulation leads to excessive nervous energy, which is discharged through behaviors recognized as expressive

26 Principle of Nervous Discharge

The *Principle of nervous discharge* is related to the expression of feelings that comes directly from a build-up to the nervous system, which causes a discharge of the excitement, such as in the case of foot and finger tapping. Darwin noted that many animals rarely make noises, even when in pain, but under extreme circumstances they vocalize in response to pain and fear.

27 Progress in Emotional Computing

- Human Computer Interaction
- Computers and Robots improve services the user desires
- Computers and Robots become more than data machines
- Machines to become allied operative systems as socially sagacious factors
    — i.e Microsoft Oxford API

The progress achieved in emotional computing has not only assisted it's own research field, but also benefits practical domains such as computer interaction. Emotional computing research aims at generating computers that offer improved services that users actually desire and thus they become more than data machines. Moreover, machines are perceived as socially sagacious factors. Microsoft Oxford API (https://azure.microsoft.com/en-us/services/cognitive-services/)

is an example of the evolution of the role of machines. It is a cloud service providing recognition of emotions on the basis of facial expressions. This recognition system is based on Ekman's facial/muscle movements descriptions of emotions.

28 HCI Norms (I)

Humans can be deeply influenced by affection during social interaction. Specifically, emotional cues from others can be a powerful way to persuade people to modify their behaviours

- Influence of embodiment and Substrate of Social Robots affect user's decision making and attitude
- Humans who interact with robots with physical embodiment show significantly higher:
    - faith in them,
    - attachment
    - credibility
- Humans are able to create emotional attachment with a social robot based not only on their long-term common activities, but also on the quality of those activities

There are some evidences on how people prefer to interact with intelligent machines that can be used in the development of efficient CPS. Influence of embodiment and Substrate of Social Robots allow them to affect user's decision making and attitude [15]. Apart from that, humans who interact with robots with physical embodiment show significantly higher faith in them, attachment and credibility. Humans create emotional attachment with a social device/ robot based not only on their long-term common activities, but also on the quality of those activities [16].

29 HCI Norms (II)

- Empathy leads to acceptance or appreciation
- A personalized experience in the human- machine interaction, make people feel more comfortable with the device while they seem to have feelings of trust for that specific device.
- The insurance of confidentiality leads to greater disclosure
- Difficult to trust machines after a mistake

How can we use those theories for modeling human behavior?

29 HCI Norms (II)

A personalized experience in the human-machine interaction, make people feel more comfortable with the device while they seem to have feelings of trust for that specific device [17]. Apart from its intelligence, robot's personality is the major aspect that leads to the development of emotional attachment in human- social robot interaction [18]. Empathy leads to acceptance or appreciation and especially when it comes to robots, the intelligent social robots are able to make people feel that that they interact with a 'social' person and it is easier for them to perceive them us friends and social partners [19].The insurance of confidentiality leads to greater disclosure and so people are able to trust machines and use them more if they consider them secure [20]. On the other hand, it is difficult to trust a machine again after it makes a mistake [21].

But how can we use those theories for modeling human behavior?

30 How to use theories for Modeling Human Behavior

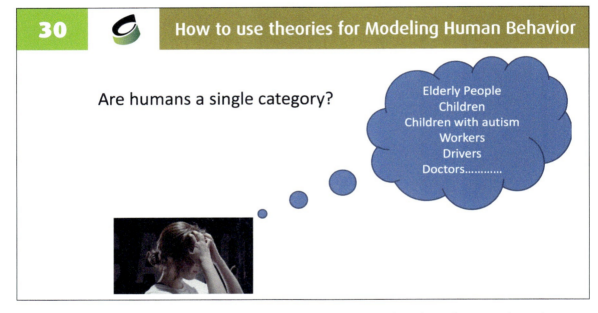

Humans belong to different categories : They have different characteristics and so they can be part of a group according to their age, profession, origin country, hobbies, etc. This is very important since there are different perceptions and abilities among a group of elderly people than in a group of children and different habits among doctors, teachers or industrial workers. Since the CPS applications are very often designed for a specific purpose, first of all we have to analyze the characteristics of the group that is going to use and interact with the CPS. In order to make sure that an effective interface is designed, it is important to keep all the principles of Human-Computer Interaction (HCI) in consideration. For example, Mahdi et his colleagues after a literature review found that the more efficient e-learning CPS, is the one meeting the requirements of the respective users [22]

Behavioural modelling is the creation of rules for the behaviour of a specific group of people. By modelling human behaviours, we improve human-machine systems. If the system understands human behaviour or anticipate it, it can adapt itself to better serve the human's needs. In order to accomplish this, the system needs to determine the human's internal state and predict states transitions to achieve the best overall efficiency[23], [24].

31 Human Behavior Modeling- Methods

- **Qualitative data**
  - Interviews
  - Focus Groups
  - Open Questions
  - Participatory Design
- **Quantitative data**
  - Questionnaires
  - Cognitive Measures
    - Memory
    - Attention

Strongly Disagree	Disagree	Undecided	Agree	Strongly Agree
(1)	(2)	(3)	(4)	(5)

There are many ways to collect data in order to model human behavior. The qualitative data, such as interviews and focus groups can give a clear idea about a specific group's behaviors. For example, if we are interested in modeling the professional drivers behavior, we can directly ask them individually or in groups about their habits. In such a way we can get significant information about aspect of their behavior with a non-numerical nature. One problem is that those who are conducting the interviews can unintentionally affect and/or manipulate the answers. Open questions in questionnaires are solving the problem of the unintentional manipulation and are better recorded and more structured.

Another qualitative approach that is the state of the art in human- machine interaction is the Participatory design, where the group of final users collaborate with the scientists in order to develop all together the application that is indented to be used by them. Concerning the social science demand for participatory design and the fact that the participatory design and all collaborative practices aim to respond better to the needs of the users, it could be considered as an approach which is more focused on the processes and the procedures of the design and not the design (result) itself. The above is especially suitable for cases that the user do not have any relative previous experience. There are three stages in the participatory design method: 1) Initial exploration of work, 2) Discovery processes, 3) Prototyping. [25], [26],

On the other hand, there are the quantitative data, more frequently extracted by likert scale questionnaires, where participants evaluate how much they agree with a specific statement from Strongly Disagree to Strongly Agree. Although it is easier to analyze those data because of their numerical nature, it needs a deep literature review about the topic before giving the questionnaires to the target group; the questions must be clear and not misleading because the participants are not able to describe their way of thinking or ask for explanations and so those who collect the data must be sure that the participants understand the questions and answer to what they are asked for. However, all those answers are self-referential. Despite the fact that most questionnaires are anonymous, participants want to show a better version of themselves –a problem that is more intense in qualitative methods- or they may not have an accurate opinion about their feelings or attitudes.

The more accurate way to measure human behavior is by measuring physical reactions and cognitive processes such as memory and attention.

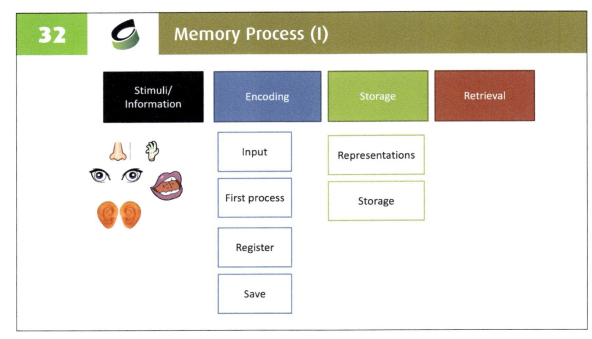

In order to use memory as an evaluation factor we first need to understand how it works. Initially, we have a stimuli or information that comes from the senses, for example a smell, image, word or touch. When those information fall into our *perception* system, we are analyzing the information in the second stage of *Encoding*. The input pass through a first process in order to understand the nature of this information (physical or chemical) and then it is registered and saved by going to the third state, the *Storage,* which results in a permanent record of the encoded information. In the storage we create representations based on the initial information. Those representations are brief descriptions of the information that help us keeping them in the storage and *retrieving* them (i.e. recalling or recognizing them) from the storage in response to certain cues in a process or activity.

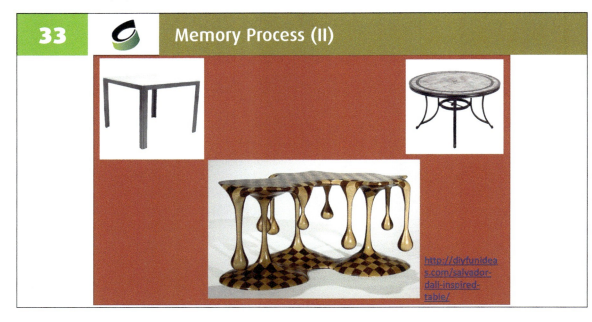

33 Memory Process (II)

Here we have an example of how the human brain creates representations about a specific object. By viewing those pictures people have a representation that they can use to characterize those objects as tables. A table can be a flat surface with 4 legs but the brain based on other information that have been stored can categorize all those images as tables (even the last one).

34 Types of Memory

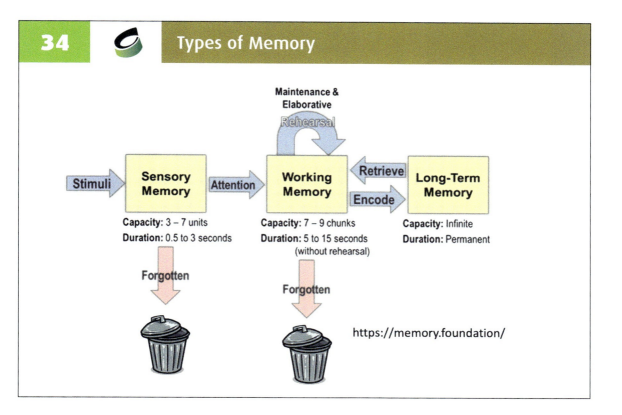

The three main forms of memory storage are sensory memory, short-term memory, and long-term memory.

Sensory memory is not consciously controlled; it allows individuals to retain impressions of sensory information after the original stimulus has ceased. It has a capacity of 3-7 units and without paying attention, the information will be deleted in approximately 0.5 to 3 seconds. By paying attention, the information will be passed to the short term memory which lasts for 5 to 15 seconds and can only hold 7 +/- 2 pieces of information at once. By using techniques such as rehearsal the information is going to the Long-term memory. An example of rehearsal is when someone gives you a phone number verbally and you say it to yourself repeatedly until you can write it down. If someone interrupts your rehearsal by asking a question, you can easily forget the number, since it is only being held in your short-term memory.

In the long term memory, the storage can hold infinite amount of information which can last for a very long time.

35 Human Behavior Modeling- Physical Reactions

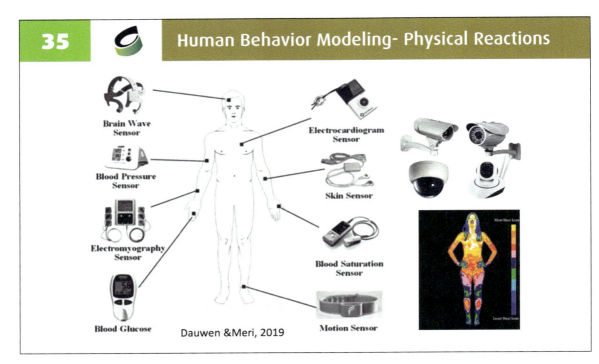

Dauwen &Meri, 2019

The human physical reactions can be measured with plenty of sensors such as brain wave sensor, blood pressure sensors, electromyography sensor, thermal cameras etc. The most important aspect when measuring the human body physical reaction by sensors is to know what triggered the measurements that we are getting. The more accurate approach for doing that is by measuring the reactions under specific tasks when also having control of the reactions of the participants.

36 Human Behavior Modeling-Scenarios

- How can we create a Cyber Physical System that will enhance students attention and memory by altering the lighting conditions?
- Transformable Lighting will improve cognitive functions

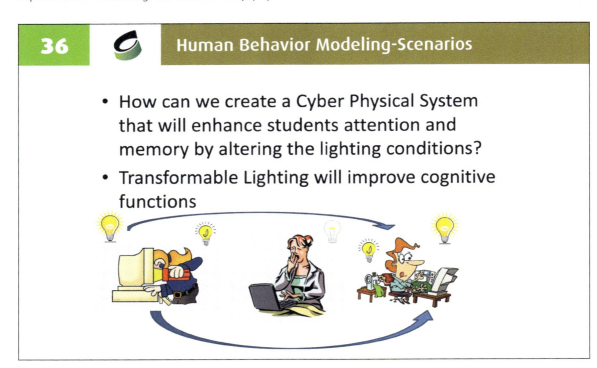

36 Human Behavior Modeling-Scenarios

How can we design specific tasks in order to be sure that we measure exactly what we want to?
Let's have an example of designing a scenario for human behavior modeling [34].

We are interested in designing an intelligent classroom that will enhance students attention and memory by altering the lighting conditions. Based on the literature review that follows, evidences show that transformable lighting conditions will improve cognitive functions (this is our hypothesis that we will try to prove).

37 Related Work on Lighting Conditions

- Circadian Rhythms
- Fluorescent light: Sustained attention and executive functions
- Intense light : improve alertness & vitality
- High luminance: discomfort in computer and paper-reading tasks
- Changing lighting conditions : improve efficient in worker's tasks, efficiency and alertness in workplaces

Environmental lighting correlates with human health and performance. Circadian rhythms are disrupted by inappropriate lighting conditions and as a result the health and performance in everyday life may be disrupted as well. [28]

Many researchers support the hypothesis that working under fluorescent light improve the executive functions and sustained attention [29], while working under intense light improve alertness and vitality [30]. On the other hand, despite the improvement in their performance, participants who were doing computer and paper-reading tasks under high luminance conditions mentioned feelings of discomfort [31]. Certain researchers varied the lighting conditions in a workplace while others allowed industry workers to change the lighting conditions in their office. Results shown significant improvement in worker's tasks and improvement of efficiency and alertness in workplaces [33].

38 — Using Memory theories for HCI

Compare Different Lighting Conditions
Measurements:
- Attention level
  - Memory test
    - Correct answers
    - Reaction time
- Long term Memory
- Short Term Memory
- Executive Functions

•A.M., Velentza, A., Nikitakis, K., Oungrinis & E., Economou. "Transformable Lighting Conditions in Learning VR Environments". In 10th IEEE International Conference on Information, Intelligence, Systems and Applications (IISA 2019), 2019, IEEE
•A. Nikitakis, A.M. Velentza, K. Oungrinis and E. Economou, "Sustained attention in smart transformable environments", 6th International Conference on Spatial Cognition (ICSC), Rome, Italy, September 2015

In order to test our hypothesis [34], we need to create a simulation of a classroom and compare different lighting conditions under specific tasks. We are interested in a variety of cognitive functions; attention, memory (short term and long term), executive functions, since all of them are vital for the learning process. Based on the way that we previously analyzed how memory works, for measuring the attention level we will use a short term memory test and compare the correct answers of the participants doing the experiment under different lighting conditions and also their reaction time (i.e. the time that they needed to reply to a question); the faster the reaction, the better the encoding and so the retrieval.

39 — Task-Procedure (I)

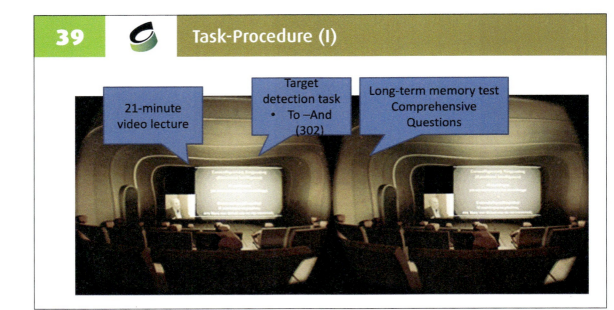

39 Task-Procedure (I)

The experiment [33], utilizes an oculus rift VR device and the participants perform a long term memory and an executive function task. They watch a 21-minute video lecture and they are tested in a target detection task and a long-term memory test based on the lecture's content. As for the target detection task, every time the speaker says the target-word, they have to press a specific button. The first target-word is the word "and" and the second is word "to". The total number of occurrences of the target-words in the lecture is 302. Participants have to press these buttons only when they hear the target-words. After the end of the lecture they answer comprehensive questions.

40 Task-Procedure (II)

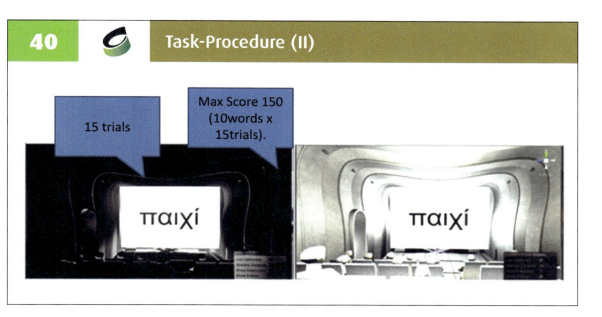

As for the sustained attention we use another task that involves both short term memory and executive functions which is a recognition task of pseudo words as analytically described in the next slide.

41 Task Example

- Example of the recognition task

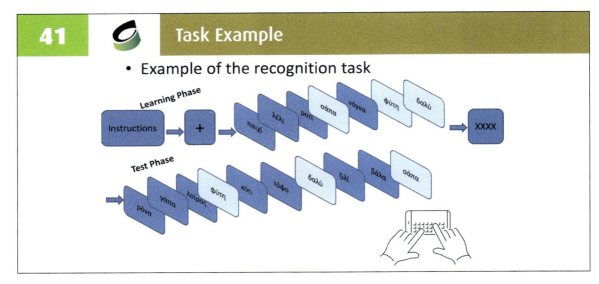

Seven pseudo words are presented one at a time for 800ms each. Subsequently participants perform a recognition task where they pick the stimuli they have seen from a new list with 3 target and 7 distractor words. Fifteen total trials are run for each participant and a correct recognition score is calculated from all trials. The maximum achievable score is 150 (10words x 15trials). At the beginning of every trial, there is a slide with instructions for 15 sec and after that it appears a slide with the symbol "+" for 800ms to prepare for memorizing the seven words that will follow. The seven pseudowords appears one after the other and each one stays in the screen for 800ms. After the seventh word, in the screen appears a mask [xxxxx] for 5 seconds so as to eliminate the effect in sensor memory. When the examination part begins, the individual has 2 sec in order to answer if the word in the screen is in the learning list or not. If it is, he/she pushes a certain button and if not another one. If the time limit of two seconds for each word passes without any button pressed, the slide changes automatically to the next word. The same procedure is repeated for 15 trials for every individual.

42 Results

- Attention- Short Term Memory Task

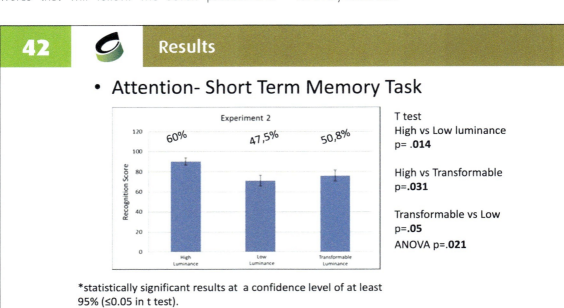

T test
High vs Low luminance
p= .014

High vs Transformable
p=.031

Transformable vs Low
p=.05
ANOVA p=.021

*statistically significant results at a confidence level of at least 95% (≤0.05 in t test).

42 Results

The main outcome of the study is that in in the transformable luminance condition participants have better scores in the executive task in comparison with both high and low lighting conditions. On the other hand, participants have higher scores in the long term memory task during the low luminance condition (in comparison with the other two conditions), while they have more correct answers in the short term memory test in the high luminance condition (in comparison with the other two conditions). In the short term memory task participants have significantly better scores in comparison with the transformable condition (at a confidence level of 97%) but also during the transformable lighting condition participants have significantly higher number of correct answers in comparison with the low luminance condition (at a confidence level of 95%). The transformable condition's participants have higher scores in the short-term memory task than those in the low luminance condition. When we compare the high luminance condition and the transformable one, the scores of the students are better in the high luminance one; however we believe that in a task with higher duration, the students exposed to transformable lighting conditions may probably have better scores because those who will be under high luminance conditions is very likely to feel discomfort. Transformable luminance may combine the best of both worlds; enhance long term memory when the students are exposed to low luminance and short term memory and sustained attention when they are exposed to high-luminance, while at the same time they will not feel any discomfort [34].

43 Use of the Results

- **Correlate data from the task with physical reactions**
- **Find when participants loose their attention**
  - Time period
  - Specific body signals
- **Design the corresponding CPS**

Back to engineering

How we can take advantage of those results in the the design of an intelligent classroom CPS? First of all it is necessary to correlate the data with the participants physical reactions. If the CPS process EEG signals it is possible to identify when participants are giving correct or wrong, slow or fast answers/reactions. By finding the spot where participants are loosing their attention we can choose the lighting conditions accordingly.

44 Discussions

- Studies in Emotions for better Human-Machine Interaction
- The role of emotion & cognition in CPS
- Cognitive Functions is a powerful weapon to design and evaluate CPS
- Engineers and Psychologists have to work together

Many theories that we are using or we are about to use in Engineering and Computer Science have been already studied, probably from a different perspective, by Psychologists and Neuroscientists. Studies in Emotion, Brain function and human behavior can/should be used for the design of efficient human-centric CPS' By measuring the emotional state and the cognitive functions we can design, test, calibrate and evaluate a CPS. Cognitive measures such as memory and attention can also give validity and reliability to our studies since we can label our data based on specific scenarios and being sure that we indeed measure the effect that we are planning to measure and also that we have the same effect when we repeat the experiment/measurements. Last but not least, Engineers and Psychologist have to work together in order to put the humans in the center of the CPS.

45 Bibliography

1. Global Cyber Physical System Market Research Report 2019 Know Market Dynamics, Opportunities and Risks 2025 Published: May 23, 2019 4:58 a.m. ET in *Market Watch*

2. Paul Korzeniowski | Feb 2018, Internet of Things (IoT) Technologies for Process Manufacturing: Global Markets, Published - Feb 2018| Code - IFT145A| Publisher - BCC Publishing

3. Gil, Miriam, et al. "Designing human-in-the-loop autonomous Cyber-Physical Systems." *International Journal of Human-Computer Studies* 130 (2019): 21-39.

4. Munir, Sirajum, et al. "Cyber physical system challenges for human-in-the-loop control." *Presented as part of the 8th International Workshop on Feedback Computing.* 2013.)

5. Liu, Zhong, et al. "Cyber-physical-social systems for command and control." *IEEE Intelligent Systems* 26.4 (2011): 92-96.

6. Miller, David Bryan, and Wendy Ju. "Joint cognition in automated driving: Combining human and machine intelligence to address novel problems." *2015 AAAI Spring Symposium Series*. 2015.

7. Castillo Aguilar, Juan, et al. "Robust road condition detection system using in-vehicle standard sensors." *Sensors* 15.12 (2015): 32056-32078.

8. Foen, N. "Exploring the Human-Car Bond through an Affective Intelligent Driving Agent". Master's thesis. MIT, Cambridge, MA (2012)

9. K. Williams, et al. "Affective Robot Influence on Driver Adherence to Safety, Cognitive Load Reduction and Sociability". In Proceedings of the 6th International Conference on Automotive User Interfaces and Interactive Vehicular Applications (pp. 1-8). ACM. 2014

10. J. Yang, et al. "Affective interaction with a companion robot in an interactive driving assistant system". In Intelligent Vehicles Symposium (IV), 2013 IEEE (pp. 1392-1397). IEEE.

11. https://en.wikipedia.org/wiki/Inside_Out_(2015_film)

12. P. Ekman, P. "Universals and Cultural Differences in Facial Expressions of Emotions". In Cole, J. (Ed.), *Nebraska Symposium on Motivation* (pp. 207-282). Lincoln, NB: University of Nebraska Press.1972

13. P. Ekman, P. & W. V. Friesen. "Constants Across Cultures in the Face and Emotion". *Journal of Personality and Social Psychology, 17(2)* , 124-129.1971

14. S. Turabzadeh et al. Facial Expression Emotion Detection for Real-Time Embedded Systems, *Technologies* 6.1: 17. 2018

15. B. Wang, B., & P. L. P. Rau. "Influence of Embodiment and Substrate of Social Robots on Users" Decision-Making and Attitude. *International Journal of Social Robotics*, 1-11. 2018

16. M. Dziergwa, et al. "Long-term cohabitation with a social robot: A case study of the influence of human attachment patterns". *International Journal of Social Robotics*, *10*(1), 163-176. 2018

17. C. M. Karat, et al. (Eds.). *"Designing personalized user experiences in eCommerce"* (Vol. 5). Springer Science & Business Media. 2004

18. L. Robert. "Personality in the human robot interaction literature: A review and brief critique". In Robert, LP (2018). Personality in the Human Robot Interaction Literature: A Review and Brief Critique, Proceedings of the 24th Americas Conference on Information Systems, Aug (pp. 16-18) 2018

19. M. Desai. "Module 4 Emotional Intelligence". In Introduction to Rights-based Direct Practice with Children (pp. 99-128). Springer, Singapore 2018

20. K. J. Corcoran. "The relationship of interpersonal trust to self-disclosure when confidentiality is assured". *The Journal of psychology*, *122*(2), 193-195.1988

21. S.Ye et al. "Human Trust After Robot Mistakes: Study of the Effects of Different Forms of Robot Communication". *International Conference on Robot & Human Interactive Communication (RoMan), 2019.*

22. Z. Al Mahdi et al. "Analyzing the Role of Human Computer Interaction Principles for E-Learning Solution Design". In *Smart Technologies and Innovation for a Sustainable Future* (pp. 41-44). Springer, Cham.2019

23. I. Oppenheim, & D. Shina. "A context-sensitive model of driving behaviour and its implications for in-vehicle safety systems". Cogn Tech Work., 13(4), 229-241. 2012

24. J. Kamla, et al. "Analysing truck harsh braking incidents to study roundabout accident risk". *Accident Analysis & Prevention*, *122*, 365-377 2019

25. C. Spinuzzi, Clay. "The Methodology of Participatory Design". Technical Communication. 52. 163-174. 2005

26. Björling, Elin A., and Emma Rose. "Participatory research principles in human-centered design: engaging teens in the co-design of a social robot." *Multimodal Technologies and Interaction* 3.1 (2019): 8.

27. M. Dauwed & A. Meri, A. "IOT Service Utilisation in Healthcare". In *IoT and Smart Home Automation*. IntechOpen. 2019

28. L. Bellia, F. Bisegna & G. Spada. "Lighting in indoor environments: Visual and non-visual effects of light sources with different spectral power distributions". Building & Environment, 46(10), 1984-1992, 2011

29. S. L., Chellappa, et al. "Non-Visual Effects of Light on Melatonin, Alertness and Cognitive Performance: Can Blue-Enriched Light Keep Us Alert?".Plos ONE, 6(1), 1-11, 2011

30. K. Smolders, Y. de Kort, & P. Cluitmans. "A higher illuminance induces alertness even during office hours: Findings on subjective measures, task performance and heart rate measures". Physiology & Behavior, 107(1), 7-16, 2012

31. L. Ji-Hyun, M., Jin Woo & K. Sooyoung. "Analysis of Occupants' Visual Perception to Refine Indoor Lighting Environment for Office Tasks".Energies (19961073), 7(7), 4116-4139, 2014

32. H. Juslén, M.,Woutersb,. & A. Tenner. "The influence of controllable task-lighting on productivity: a field study in a factory". Applied Ergonomics, 38 ,39–44, 2007

33. H.Juslén, & A. Tenner. "Mechanisms involved in enhancing human performance by changing the lighting in the industrial workplace". Int.J. Ind. Ergonomics 35 (9), 843–855, 2005

34. A.M., Velentza, A., Nikitakis, K., Oungrinis & E., Economou. "Transformable Lighting Conditions in Learning VR Environments". In 10th IEEE International Conference on Information, Intelligence, Systems and Applications (IISA 2019), 2019, IEEE

35. A. Nikitakis, A.M. Velentza, K. Oungrinis and E. Economou, "Sustained attention in smart transformable environments", 6th International Conference on Spatial Cognition (ICSC), Rome, Italy, September 2015

36. T. Jones & K. Oberauer. "Serial-position effects for items and relations in short-term memory". Memory, 21(3), 347-365, 2013

CHAPTER 06

Analog IC Design for Smart Applications in a Smart World

Prof. Georges Gielen

KU Leuven, Belgium

#		Page	#		Page
1.	Introduction	214	16.	Capacitive Time-based Sensor Interface	224
2.	Outline	214	17.	Resistive Time-based Sensor Interface	224
3.	Sensing is Ubiquitous	214	18.	Use of Emerging Nanotechnologies	225
4.	IoT Market Forecasts	215	19.	Example: Sensing With Embedded IR	226
5.	Requirements for Sensing	216	20.	How to Achieve Ultra-miniaturization?	226
6.	Example: Neural Probe	217	21.	Reliability/robustness Considerations	227
7.	455-Activ-Electrode Neural Probe	217	22.	Robust 0.18μm CMOS Readout Chip	228
8.	Outline	218	23.	Self-healing Analog Chip	228
9.	Further Miniaturization Possible?	219	24.	Outline	229
10.	Healthcare: Treatment of Inflammation	219	25.	Smart System Architecture	230
11.	Example Implementation: StimDust	220	26.	Compressed sensing	230
12.	What Makes Analog Circuits Large?	221	27.	Signal-dependent Processing	231
13.	Towards Digital-only Analog Circuits	222	28.	Event-based Sensing	232
14.	Time-based Architecture (I)	222	29.	Conclusions	232
15.	Time-based Architecture (II)	223	30.	Acknowledgment	233

Chapter six presents some challenges for the design of integrated circuits (ICs) for the "smart society", where an increasing number of applications provide intelligent services that will make our daily lives more comfortable and with unprecedented functionality. Sensing is used in all applications where the physical world interacts with the electronic world. This happens in almost all application fields, where physical signals are converted into electronic signals, and back. Networked sensing has become ubiquitous in today's smart world, with Internet of Things, personalized medicine, autonomous vehicles, etc. as main application drivers. After the short introduction about the ubiquitous need for sensing in a smart world, it is discussed how to make the sensing readout circuits small, power efficient and smart. This can be realized by adopting highly-digital time-based architectures, resulting in technology-scalable circuits with a small chip area. Most applications also require an extremely high reliability and resilience of the sensing devices against all possible variations (production variability, temperature, supply, EMI, etc.) as well as against degradation and drift over time. To deal with the increasing amounts of data, modern applications require more intelligent computing in the edge. This can be achieved by performing information rather than data extraction in the edge, by designing circuits that are fully context adaptive and that preferably have built-in learning capability. Several successful real CMOS designs are presented to illustrate these concepts.

1 Introduction

Our world is moving towards a "smart society" where an increasing number of applications provide intelligent services that will make our daily lives more comfortable and with unprecedented functionality. This presentation will describe some challenges for the design of integrated circuits (ICs) for these applications. This will be illustrated with some of the many IC designs that we carry out annually as part of the micro/nanoelectronics circuit research at KU Leuven university in Leuven, Belgium. The presenter, Prof. Georges Gielen, can be contacted at gielen@kuleuven.be.

2 Outline

- **ubiquitous sensing in a smart world**
- how to make sensing small
- how to make sensing smart
- conclusions

This is the outline of the presentation. After a short introduction about the ubiquitous presence of sensing in a smart world, we will focus on two aspects : 1) how to make the sensing circuits small, and 2) how to make the sensing circuits smart. We will finish with some conclusions.

3 Sensing is Ubiquitous

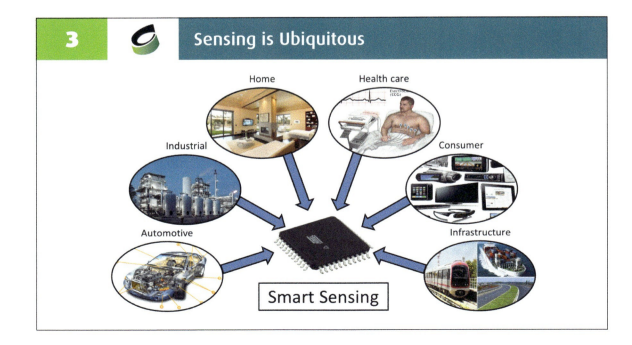

ANALOG IC DESIGN FOR SMART APPLICATIONS IN A SMART WORLD

3 Sensing is Ubiquitous

Sensing is used in all applications where the physical world interacts with the electronic world. This happens in almost all application fields, ranging from communication, automotive and industrial over infrastructure and domotic to consumer and healthcare applications. They all interact with the real world through sensing, where physical signals are converted into electronic signals, and back. The coupling to electronic processing allows to make many applications smart or intelligent, i.e. that they selectively can adapt and respond to the actual situation.

One example is smart farming, that can for instance result in higher crop yields with less spraying.

4 IoT Market Forecasts

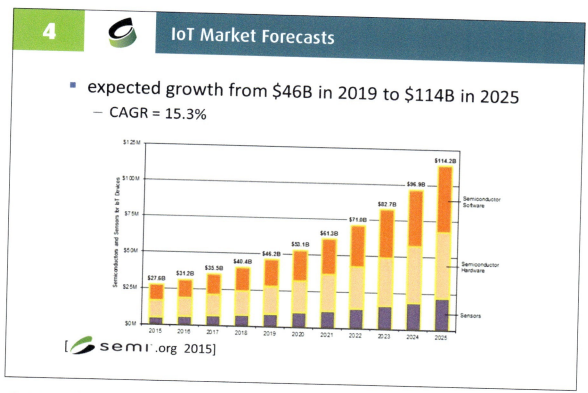

An interesting evolution of the past years is that sensing devices are networked and connected to the internet or the "cloud", often in a wireless way. This results in the so-called Internet of Things (IoT), where billions of devices are connected through the internet. The figure on the slide, composed by the Semi organization, shows the evolution of (forecasted) IoT market figures [semi.org 2015]. (Note that many other market forecast figures can be found on the web). According to this figure, from 2019 till 2025 the IoT market is forecasted to grow from 46B$ to 114B$, which is an impressive CAGR of 15.3%. Compare it to the lower average growth of the entire semiconductor market ! Even if the actual numbers will be slightly off later on, it remains an impressive growth rate.

5 Requirements for Sensing

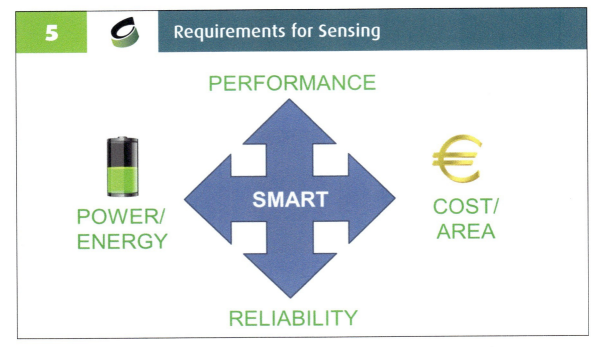

So, what are the requirements for the sensing devices in such IoT applications? Above all, they must meet the technical performance requirements for the application at hand. Secondly, they must have a low energy consumption, especially in applications powered from a limited energy source like a battery or an energy harvester. This implies that circuits must be power-efficient and only turned on when really needed. Thirdly, they must be cheap to be affordable in the application at hand. The allowed cost of course differs from for instance tagging devices in a supermarket or warehouse to implanted biomedical devices, to name a few. In case of an integrated chip solution for the sensing device, large part of the recurring cost depends on the technology used to fabricate the chip and on the area or size of the chip. Fourthly, most sensing applications require a high reliability and robustness against environmental variations (e.g. temperature, supply, humidity, EMI, etc.).

Safety-critical applications like autonomous cars or implanted electronics even require absolute robustness.

And finally, in recent years, there is also the requirement to make the sensing device smart and adaptive. The nice thing in the electronics industry over many decades is that the continuous downscaling of the semiconductor technology according to Moore's law helps to achieve the above requirements, both in terms of enabling extra functionality and of reducing cost and form factor.

6 Example: Neural Probe

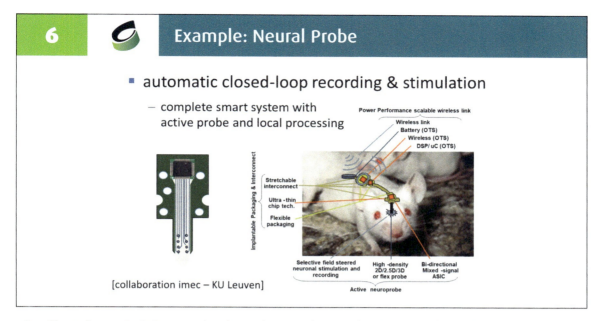

[collaboration imec – KU Leuven]

An illustrative example is a neural probe as shown in the figure left, that can be used for neural research or for medical treatments. The presented results are from a collaboration between the imec research institute and KU Leuven university. The probe's needle is inserted into a targeted area of the brain, where it (hopefully) is in close vicinity to some neurons. It has a number of electrodes that can record the signals of these neurons. These signals are converted and transmitted outside (e.g. wirelessly) by a readout chip at the base of the probe (see figure right). Some of the electrodes do not record, but are used for stimulation of the surrounding neurons.

Ultimately, for medical treatments, the goal is to build a closed-loop system that can locally record and process the signals and have the system react autonomously through the proper stimulations. Steady progress is being made toward this goal, but many technological developments and improvements are still needed.

7 455-Active-Electrode Neural Probe

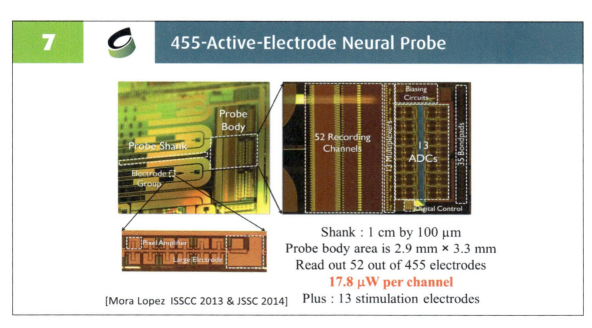

[Mora Lopez ISSCC 2013 & JSSC 2014]

Shank : 1 cm by 100 µm
Probe body area is 2.9 mm × 3.3 mm
Read out 52 out of 455 electrodes
17.8 µW per channel
Plus : 13 stimulation electrodes

7 455-Activ-Electrode Neural Probe

This slide shows photographs of a silicon-integrated implementation of such neural probe with 455 active electrodes, first presented at the ISSCC conference in 2013 by Mora Lopez et al. and later elaborated in IEEE JSSC 2014. The probe has been fabricated in a standard CMOS technology with additional postprocessing to deposit the electrodes. The probe needle or shank measures 1 x 0.01 cm², while the readout chip measures 2.9 by 3.3 mm². The probe has 455 active recording electrodes, out of which 52 can be read out in parallel at any moment in time, in addition to 13 stimulation electrodes distributed across the shank. The power consumption per readout channel is 17.8 microwatt. Recent work has improved beyond this design, integrating many more electrodes on one probe today.

8 Outline

- ubiquitous sensing in a smart world
- **how to make sensing small**
- how to make sensing smart
- conclusions

If size and cost are so important for sensing applications, how can we then make the sensing interface circuits small ? And how can we reduce the power consumption ?

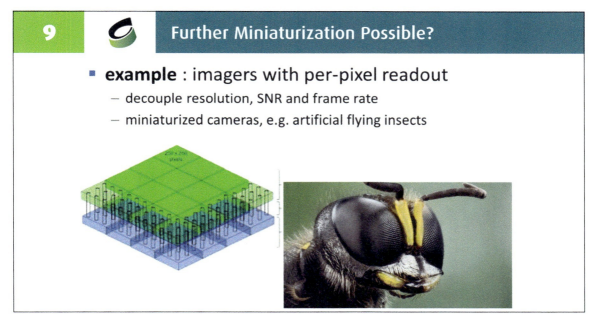

Think of what could be possible if electronics would be further miniaturized and reduced in area and size ! For example, as shown in the slide, we could design and build imagers with per-pixel readout, allowing to make ultra-miniaturized cameras for monitoring or surveillance. These could be integrated in (sun)glasses, artificial retinas for blind people, artificial flying insects, etc. Applications abound.

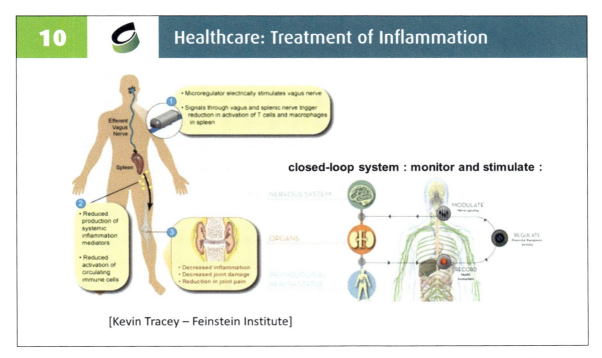

Another example is healthcare, where interest is increasing in preventive besides curative medicine, and in personalized rather than generic medicine. This requires the monitoring of biosignals on the body (e.g. monitoring of stress indication signals) and/or in the body (e.g. cancer biomarker

10 Healthcare: Treatment of Inflammation

monitoring), and adequate signal processing algorithms based on which the proper personalized actions can be initiated.

An illustrative example is shown in the slide. It was presented by Kevin Tracey from the Feinstein Institute to illustrate the novel concept of electroceuticals (the electronic equivalent of biochemical pharmaceuticals): implanted therapeutic electronic devices that stimulate the molecular mechanisms of human organs by generating electric pulses on the neural nerves coming from the brain. The slide shows the example of the treatment of patients with insufficient inflammation suppression: the presence of inflammation is detected electronically, and this information is fed to the electronic device that has been implanted around the vagus nerve; by electrically stimulating the vagus nerve, the spleen is stimulated to increase the biological immunity mechanisms inside the body to suppress the inflammation.

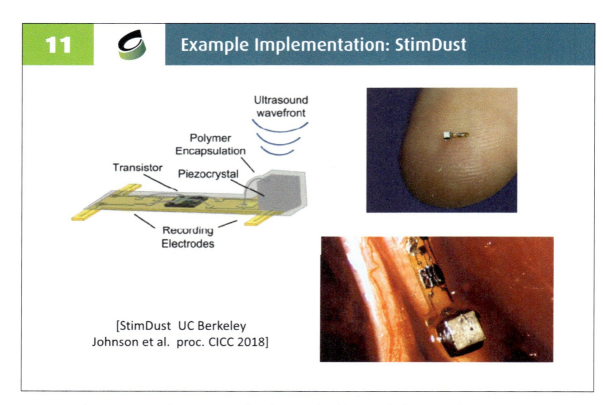

11 Example Implementation: StimDust

[StimDust UC Berkeley Johnson et al. proc. CICC 2018]

A practical prototype implementation of such in-body biomedical monitoring device is presented in the slide: the StimDust system (figure left) developed by Johnson et al. at UC Berkeley [proc. CICC 2018]. Being really tiny (top right), it can be inserted around for instance a nerve to record the local biosignals (bottom right).

Powering and (bidirectional) communication happen through ultrasound waves. Further developments to make the devices even smaller might result in true "body dust"-like monitoring devices.

12 What Makes Analog Circuits Large?

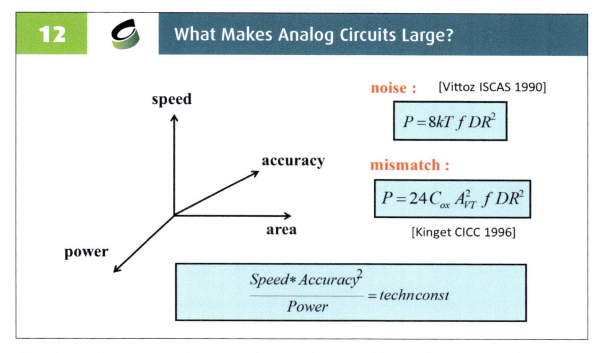

So what makes sensor interface circuits large? Since the signals in the physical world are analog in nature, sensing interfaces are typically analog/mixed-signal circuits that amplify and convert the measured signals into digital format for further processing. Being analog circuits, they are subject to the typical fundamental relationships that govern analog circuits: the power-speed-accuracy trade-off. Both because of thermal noise [Vittoz proc. ISCAS 1990] and because of device mismatch [Kinget proc. CICC 1996 & JSSCC 2005], analog circuits require a minimum power consumption to process signals at a given speed and accuracy (dynamic range), as shown in the two equations top right.

Since both equations have a very similar shape, they can be generalized to the equation in the bottom middle: speed times accuracy² over power is constant. Area is the binding factor that connects these characteristics. For instance, to increase speed, devices can be made smaller, but this would reduce the accuracy. To keep the accuracy and get higher speed at the same time, the devices cannot be made smaller, but the power needs to be increased.

In many analog circuits in practice, the mismatch is the actual limiting factor to the power consumption. Therefore, since mismatch is not improving much with technology, traditional analog circuits are not really shrinking in area nor power consumption with technology scaling. This problem is even worsened by the reduction in voltage headroom caused by the shrinking supply voltages.

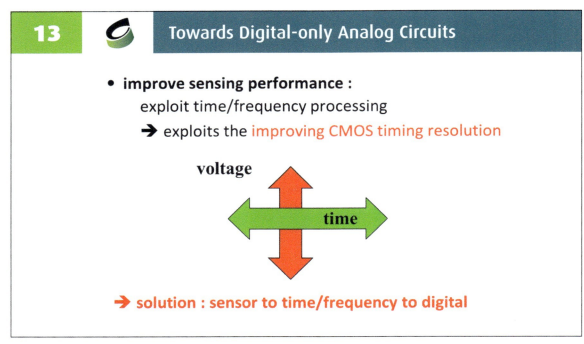

So how can we then make sensing interface circuits that are small in area? While the dynamic range may be difficult to achieve in amplitude swing in advanced CMOS technologies, especially due to the lower supply voltages, the resolution in time does improve continuously in scaled processes. The transistors switch faster. Therefore, processing analog signals in time in so-called time-based or time-encoding analog circuits may be a more scalable alternative. The principle is then simple: the signal being sensed is converted into a time or frequency signal which is then digitized directly. As we will see, this principle can be implemented with almost only digital circuits. Therefore, these analog circuits are implemented as highly-digital circuits that are technology scalable.

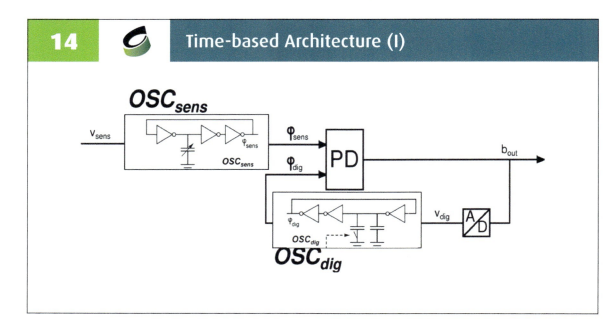

14 Time-based Architecture (I)

So how do we convert an analog input sensor signal into time or frequency ? A highly-digital way to do this is to use a ring oscillator. Assume that the sensor is capacitive, which means that the sensor capacitance varies with the input signal to be sensed. By inserting the capacitance as load to a stage in the ring oscillator (labeled osc_sens in the slide), the frequency of oscillation varies with the input signal v_sens.

So how do we then read it out as digital output ? By using a second identical ring oscillator (labeled osc_dig) and by locking the two ring oscillators in phase with a phase detector (labeled PD). The frequency of this second oscillator is controllable by means of a variable capacitive load that can be switched in or out, steered by the digital output signal bout coupled back. Since the feedback loop makes both oscillators lock in phase, in equilibrium they run at the same frequency. Therefore, the digital output signal is the digital representation of the analog sensor input signal.

[This is published by Van Rethy et al. in Microelectronics Journal 2014]

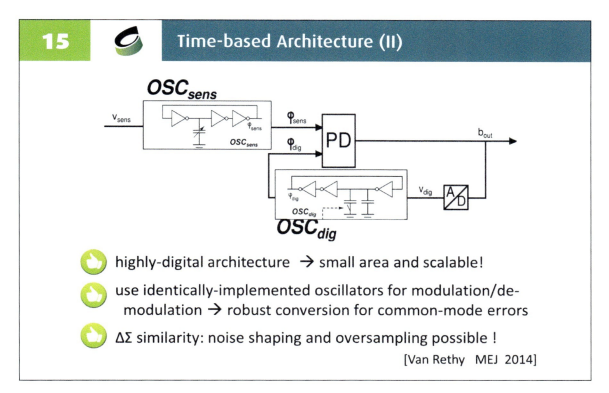

The presented time-based architecture has some interesting properties [Van Rethy et al. Microelectronics Journal 2014] :
1) all circuits used (ring oscillators, phase detector) are highly digital, resulting in a technology-scalable solution with low area;
2) since the architecture depends on the phase locking of two identical oscillators, it is highly insensitive to common-mode errors that affect both oscillators in the same way;
3) the structure is similar to a Delta-Sigma modulator but in the phase domain --> since the frequency to phase conversion of the oscillators serves as integrator, oversampling will result in first-order noise shaping!

Capacitive Time-based Sensor Interface

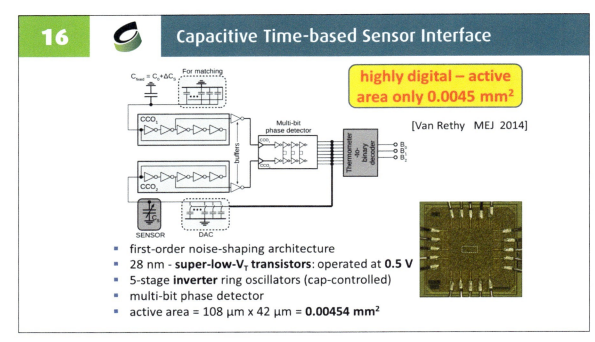

highly digital – active area only 0.0045 mm²

[Van Rethy MEJ 2014]

- first-order noise-shaping architecture
- 28 nm - **super-low-V_T transistors**: operated at **0.5 V**
- 5-stage **inverter** ring oscillators (cap-controlled)
- multi-bit phase detector
- active area = 108 µm x 42 µm = **0.00454 mm²**

The chip presented on this slide is an example of a time-based sensor readout circuit for a capacitance sensor. Different from the previous slide, one oscillator has a fixed capacitance as load, while the oscillator in the feedback loop contains both the variable sensor capacitance and the digitally controlled capacitance. The advantage of this alternative architecture is that the oscillators operate at a constant frequency, hence the circuit's performance is not hampered by the nonlinear conversion characteristic of the oscillators. The structure has a multi-bit phase detector in this case. The chip [published by Van Rethy et al. in the Microelectronics Journal in 2014] has been designed in 28 nm CMOS, is operated with a supply voltage as low as 0.5 volt, and has an active area of only 0.00454 mm².

Resistive Time-based Sensor Interface

[Van Rethy JSSC 2013]
[Van Rethy A-SSCC 2012]

- force-balance the bridge
  - 2nd-**order BBPLL** architecture
  - implicitly digitizes the sensor signal
    - sensor to digital conversion
  - high PSRR is always maintained
- small area and technology scalable !
- implemented in 130 nm at 1V

17 Resistive Time-based Sensor Interface

A similar time-based architecture can be derived for a resistive sensor. In this case the sensor resistor (or in the slide two differential resistors) is put in a Wheatstone bridge structure, together with resistors that are digitally controlled by the output signal to switch in or out extra resistance. The two oscillators (labeled the sensor VCO and the loop VCO) are steered by the two midpoints of the bridge structure. The feedback loop and the phase detector make the two oscillators operate at the same frequency, the digital output therefore balances the bridge and hence becomes the digital readout of the analog resistive sensor signal. Adding a digital PI filter after the binary phase detector turns this into a second-order architecture. The chip shown here [published by Van Rethy, Danneels et al. in IEEE JSSC 2013] has been implemented in 130nm CMOS, a typical technology for sensing applications.

18 Use of Emerging Nanotechnologies

- **Carbon NanoTube (CNT)** technology is an excellent candidate
  - CNFETs are projected to improve in **energy-delay product with >10x** compared to Si CMOS at highly-scaled nodes
- CNTs are also ideal to be functionalized as **sensors**!
- have demonstrated several interface circuits [Shulaker & Van Rethy ISSCC 2013 & JSSC 2014]

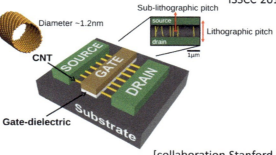

[collaboration Stanford U – KU Leuven]

While CMOS time-based architectures show a low chip area, their energy efficiency is comparable to that of classical voltage-based sensor readout architectures. Since the time-based structures are highly digital, one way to improve the energy efficiency is to implement the circuit in a technology that has a better energy-delay product. Carbon NanoTube (CNT) technology is an excellent candidate for this, since its energy-delay product is projected to be at least an order of magnitude better than CMOS at highly scaled process nodes. In a collaboration between Stanford University and KU Leuven university, several prototype structures for sensor readout have been designed and demonstrated in CNT technology by Shulaker and Van Rethy et al. [proc. ISSCC 2013 & IEEE JSSC2014].

19 Example: Sensing With Embedded IR

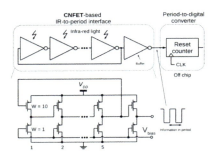
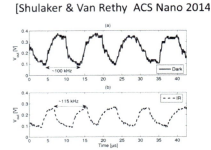

[Shulaker & Van Rethy ACS Nano 2014]

- oscillator functions as **infrared sensor**
  - inverters sensitive to infrared
- open-loop architecture :
  - 5-stage ring oscillator and reset counter to digitize
- for 32 nm CNT: **100 kHz** range, **power only 130 nW**

An extra benefit is that the CNT readout circuit itself can be functionalized to serve directly as sensor. This is illustrated in this slide for infrared (IR) sensing. Since CNTs are sensitive to IR light, a CNT circuit can directly act as IR sensor. The circuit structure is shown in the upper left of the slide, while some measured signal waveforms are shown in the upper right. The ring oscillator used in this case has 5 stages. Fabricated in an academic 32nm CNT technology, the circuit achieves a frequency of 100 kHz and consumes only 130 nW. The chip has been published by Shulaker and Van Rethy et al. in ACS Nano 2014.

20 How to Achieve Ultra-miniaturization?

- **implement digital-like where possible**
  - avoid analog to implement analog
- **exploit time / oversampling / averaging...**
- **use advanced nanotechnologies**
  - CMOS, CNT...
  - use the 3rd dimension : 3D stacking
- **exploit the system level**
  - redundancy / sensor fusion
  - exploit signal characteristics / information features...
- ...

20 How to Achieve Ultra-miniaturization?

In summary, here are some guidelines to achieve high area miniaturization for analog circuits:
1) Avoid analog circuits to implement analog functions --> implement the analog functionality with digital-like circuits where possible;
2) exploit the high resolution in time of advanced CMOS – use oversampling, averaging, etc. especially for applications that have a low bandwidth;
3) using advanced nanotechnologies beyond CMOS can help too, e.g. CNT, 3D stacking, etc.;
4) exploit the system level: by using sensor redundancy, sensor fusion, exploiting information features and signatures, etc., an accurate sensing system can be built, even if the individual sensor readouts are less accurate... (very similar to what our brain does)

21 Reliability/robustness Considerations

- sensor and sensor interface need to be designed for **high accuracy, VDD/temperature resilience, low power and low area**   `EXTENDED TEMPERATURE RANGE: -40°C to 175°C`
- low (phase) noise and use of oversampling
- mitigate effect of mismatch → time-based chopping

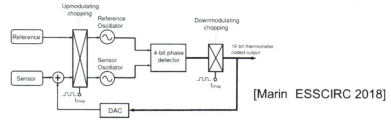

[Marin ESSCIRC 2018]

- **compensate for drift, time variations, aging, etc.**

An important constraint for many sensing applications are the strong requirements in terms of reliability and robustness. The circuit must function correctly even when the supply voltage shows peaks and dips, when the temperature changes (e.g. the extended temperature range for automotive ICs is from -40°C to +175°C), when electromagnetic interferences (EMI) are present, etc. In addition, the circuit must be robust against unavoidable production tolerances and mismatches ("process variability"), as well as against drift and aging-induced performance degradation with time.

Let us consider again the time-based sensor readout architecture presented before, that was relying on the matching of the two identical oscillators. To mitigate the effect of unavoidable production mismatches between the two oscillators, the technique of chopping can be applied, be it in the time domain rather than in the amplitude domain as in traditional chopping. As indicated in the figure, this means that the different paths in the circuit are switched alternatingly, in the shown architecture a) at the inputs to the oscillators and b) digitally after the phase detector. This reduces the impact of mismatches between the two oscillators.

22 Robust 0.18μm CMOS Readout Chip

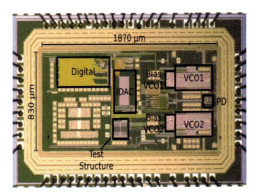

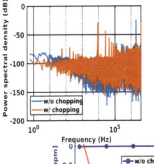

[Marin ESSCIRC 2018 & JSSC 2019]

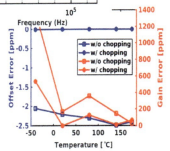

- 15.0 ENOB achieved
- extremely low drift due to combination of chopping and VCO tuning
  - over the entire temperature range from -40°C to 175°C

The time-based architecture of the previous slide has been implemented on a prototype chip in 180nm CMOS by Marin et al., as described in proc. ESSCIRC 2018 and elaborated in IEEE JSSC in 2019. Although time-based chopping is effective (as shown in the measured output spectrum in the figure top right) and results in an ENOB of 15.0 bits, analysis shows that it does not solve all issues due to fabrication mismatches between the two oscillators. Therefore, additionally a tuning step is needed to calibrate the two VCO frequencies, which can be done in a relatively simple way (see the referenced publications). Thanks to this combination of chopping and VCO tuning, the chip shows an extremely low drift across the entire automotive temperature range from -40°C to +175°C, as shown in the plot with measured results in the bottom right.

23 Self-healing Analog Chip

Another problem are the changes in performance with time, due to the aging of the devices as a result of the chip's usage. To cope with this, "self-healing" chips can be built that follow the sense-and-react paradigm: monitoring circuits are added on chip to measure the amount of aging degradation; this degradation information is sent to an on-chip controller, that then takes the necessary corrective actions to keep up performance. This has been demonstrated experimentally on a self-healing analog output driver chip by De Wit et al. [proc. ESSCIRC 2011 & IEEE JSSC 2012]: the degradation of the output transistors is monitored by measuring their current to a reference, and, when needed, the controller autonomously adds spare subtransistors to boost again the output drive capabilities. This is illustrated in the figure bottom right, where you see the chip "healing itself" at some instances in time (mimicked in the experiment by the temperature).

23 Self-healing Analog Chip

- **on-chip power efficiency monitor** [De Wit ESSCIRC 2011 & JSSC 2012]
  - measure output stage on-resistance ($R_{on}=2.25\Omega$)
  - compare current R_{on} to reference resistor R_{ref}
- **on-chip automatic controller**
  - switch in extra output transistors
- **real-time self-healing capabilities**

$$\eta = \frac{P_{load}}{P_{load}+P_{loss}} = \frac{R_l}{R_l+R_{on}}$$

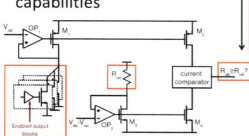
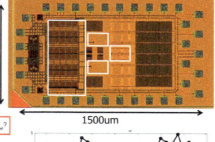
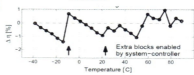

24 Outline

- ubiquitous sensing in a smart world
- how to make sensing small
- **how to make sensing smart**
- conclusions

The next question in today's context of smart systems and Internet of Things is how to make the sensing interfaces smart?

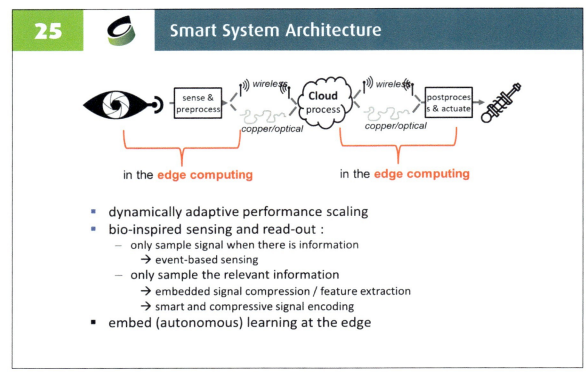

The key here is that the increasing deployment of IoT devices and connected sensors will create huge amounts of data. Transmitting those data to the cloud for further processing consumes enormous amounts of energy and drains the limited energy supplies (e.g. batteries) of the edge devices. Moreover, many of these data often do not contain really useful information. For example, in normal situations your body's stress level will probably not change much every second. Same with what you see through a camera. These arguments therefore call for more local processing or "computing in the edge", in our case in the sensing devices.

To make the interface smart, we need to be able to scale up or down its performance adaptively and dynamically, depending on the context at hand. To save power and limit data, we can also look at how nature solves these issues and adopt similar bio- inspired techniques in our circuits. This includes event-based sensing (i.e. only sampling a signal when there is a significant change) and signal compression (i.e. compressing the data or extracting information/features instead of full data), etc.

And the ultimate sign of smartness is to build some form of (autonomous) learning into the interface, so that the system can adapt itself, and learn and improve from its experiences.

Compressed sensing

One form of data compression is to apply compressed sensing. This has been investigated by several researchers. The slides show results from Prof. Atienza's group at EPFL, where compressed sensing was applied to ECG signals.

Compression resulted in less data, and even though this was only 8% of data compared with traditionally sensed data, the signal could be reconstructed with full signal fidelity. Obviously, this results in huge power savings in the edge device.

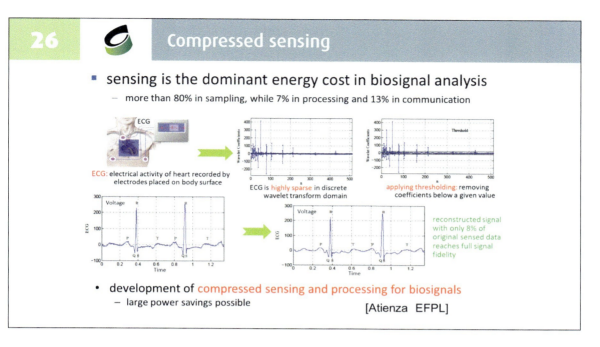

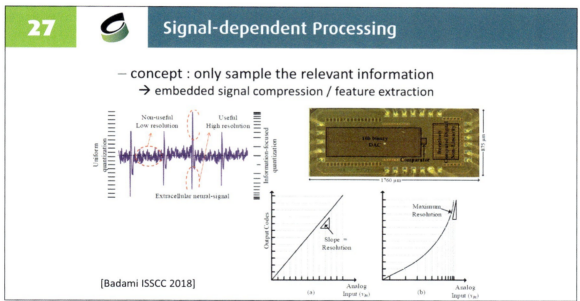

An example of signal-dependent processing has been demonstrated by Badami et al. in proc. ISSCC 2018. The application is recording neural signals, which typically consist of a noisy background signal and individual spikes from the neurons near the recording electrodes. In most cases, the information of interest is in the occurrence and shape of these spikes (see figure top left). Therefore, instead of digitizing the recorded signals with a converter with linear (= uniform – see bottom left) characteristic, it is better in this application to use a converter with nonlinear (= non- uniform) characteristic, where the resolution is low for small signal values and where the resolution is high for large signal values (i.e. the spikes). Obviously, whether this results in any benefit depends on the application; but the converter characteristic can be customized to the targeted application.

28. Event-based Sensing

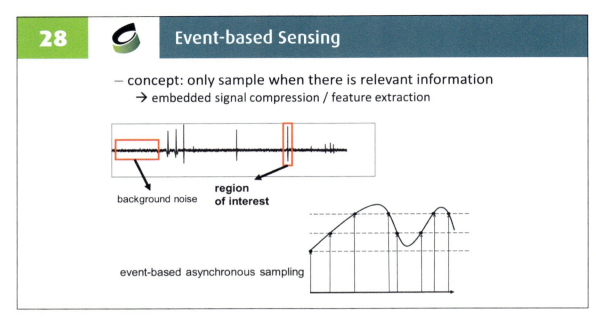

— concept: only sample when there is relevant information
→ embedded signal compression / feature extraction

background noise

region of interest

event-based asynchronous sampling

A final technique presented here is not to be adaptive in amplitude, but rather to be adaptive in time. Event-based sensing only processes the signal when it is really changing. Consider again the example of a neural signal: most of the time there is only background noise. If the interest of the application is in the spikes, then it is sufficient to build a sensing interface that is only active when there is a spike in the signal, and that is off or in some kind of low-energy sleep state otherwise. One possible technique to do this is for instance level-crossing conversion, which only samples the signal when it changes more than a LSB (i.e. crosses a level). This results in an asynchronous digital output, different from traditional fixed-rate sampling and conversion. Depending on the application, this may result in large energy savings. A typical example today are event-based cameras, but also other applications can benefit if the signals involved have low activity (or high sparsity in time).

29. Conclusions

- networked sensing has become ubiquitous
  - Internet of Things, personalized medicine, etc. as key drivers
- need for miniaturized sensor interfaces
  - highly digital architectures – small area - scalable
  - high power efficiency
- need for ultra-reliable electronics
  - extreme reliability
  - low drift – high resilience
- need for intelligent edge computing
  - embedded information extraction in the edge
  - fully adaptive processing with learning capability
  - exploit system-level redundancy with "simple" nodes

29 Conclusions

We can now summarize the conclusions from this presentation :
1) Networked sensing has become ubiquitous in today's smart world, with Internet of Things, personalized medicine, autonomous vehicles, etc. as main application drivers.
2) These applications need many small and networked sensing devices. For the sensor interfaces, this can be realized by adopting highly-digital time-based architectures, resulting in technology-scalable circuits with a small chip area and a high power efficiency
3) Most applications also require an extremely high reliability and robustness of the sensing devices against all variations (production tolerances, temperature, supply, EMI, etc.) as well as against degradation over time. Techniques to achieve this have been presented.
4) Finally, today's applications more and more require intelligent computing in the edge. The energy and data bottlenecks can be limited by performing embedded information (rather than full data) extraction in the edge, by designing circuits that are fully adaptive in amplitude and time and that preferably have built-in learning capability. Note also that high performance at system level can also be created by combining information from multiple "simple" devices, avoiding that all devices themselves must be too sophisticated.

30 Acknowledgment

- many thanks to all PhD students for their innovative research contributions

- many thanks to all collaborators from KU Leuven and other universities and companies

- many thanks to all funding organisations: KU Leuven DOC, Flemish FWO, Flemish IWT/VLAIO, EU and the many companies supporting the research

- **contact : gielen@kuleuven.be**

This presentation is based on work by many PhD researchers in my research group at KU Leuven. The presenter wants to thank them all for their innovative research contributions. Many thanks also go to all collaborators from within KU Leuven and other universities and companies. Finally, sincere thanks to the many organizations and companies that fund my research.

If you are interested in this research, would like more detailed explanations or have further questions, please contact the presenter by e-mail at **gielen@kuleuven.be.**

CHAPTER 07

Powering Cyber-Physical-System nodes by Energy Harvesting

Dr. Michail E. Kiziroglou

International Hellenic University, Greece

#	Section	Page
1.	Outline	238
2.	Demand for Energy Autonomy	238
3.	Energy Harvesting as a Local Power Source (I)	239
4.	Speed vs Efficiency in Energy Flow	239
5.	Energy Harvesting as a Local Power Source (II)	240
6.	Energy Harvesting as a Local Power Source (III)	240
7.	Power Availability	241
8.	Summary of Thoughts	241
9.	Motion Energy Harvesting (I)	242
10.	Motion Energy Harvesting (II)	242
11.	Motion Energy Harvesting (III)	243
12.	Motion Energy Harvesting: Transduction Methods	243
13.	Motion Energy Harvesting: Electrostatic Example (I)	244
14.	Motion Energy Harvesting: Electrostatic Example (II)	244
15.	Motion Energy Harvesting: Electrostatic Example (III)	245
16.	Motion Energy Harvesting: Impulse Excitation and Frequency Up-conversion	245
17.	A Rotational Motion Example: The FliteWISE system (I)	246
18.	A Rotational Motion Example: The FliteWISE system (II)	246
19.	Thermal Energy Harvesting (I)	247
20.	Thermal Energy Harvesting (II)	247
21.	Thermal Energy Harvesting (III)	248
22.	Dynamic Thermoelectric Energy Harvesting (I)	248
23.	Dynamic Thermoelectric Energy Harvesting (II)	249
24.	Dynamic Thermoelectric Energy Harvesting (III)	249
25.	A Thermoelectric Example: The StrainWISE system	250
26.	Prototypes by the Imperial / Berkeley group	250
27.	Summary	251

In this chapter the topic of powering Cyber-Physical-System nodes by Energy Harvesting is discusses with a special focus on the current state-of-art of motion and heat energy sources and various applied implementation examples. The Learning Objectives of the chapter include understanding energy harvesting as a local power source, the ability to estimate environmental power availability and to assess the viability of applications based on size, environmental conditions, power demand and operation schedule specifications. Additional learning objectives include the ability to apply existing architectures of self-powered systems, and to appreciate the main challenges and realistic anticipation of the emerging power autonomy technology. The presentation includes a discussion of energy harvesting as a local power source, an overview of power availability, examples of motion and thermal energy harvesting systems and a summary of current challenges and technology expectations.

1 Outline

Learning Objectives:
- Understand energy harvesting as a local power source
- Estimate environmental power availability
- Assess application viability based on:
    - Size
    - Environmental conditions
    - Power demand
    - Operation schedule
- Use existing architectures of self-powered systems
- Appreciate main challenges and realistic anticipation
    - of the *emerging power autonomy technology!*

Presentation outline
- Energy harvesting as a local power source
- Power availability
- Motion energy harvesting
- Thermal energy harvesting
- Power line harvesting
- Examples of power autonomous systems
- Challenges and expectations

The learning objectives of this lecture are to gain an understanding of energy harvesting as a local power source, to be able to calculate the available environmental energy and to assess the viability of power autonomy for a given application.

2 Demand for Energy Autonomy

- Wireless devices need energy portability
- Batteries require recharging / replacement
- Recharging services is just about fine for a personal device
- Currently: 2.7 devices / person
- Expected change: +1 device every couple of years
- Soon: Not enough population to cover recharging demand

Methods
- Automated recharging
- Wireless power transfer
- Energy harvesting

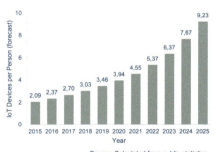

Ten devices per person globally means there is not enough man power for recharging. Possible solutions include automated recharging, wireless power transfer and energy harvesting

3 Energy Harvesting as a Local Power Source (I)

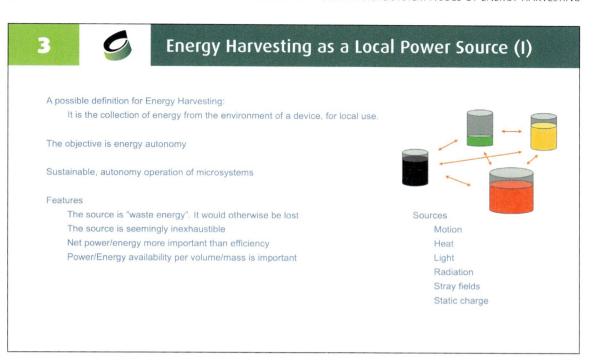

A possible definition for Energy Harvesting:
 It is the collection of energy from the environment of a device, for local use.

The objective is energy autonomy

Sustainable, autonomy operation of microsystems

Features
 The source is "waste energy". It would otherwise be lost
 The source is seemingly inexhaustible
 Net power/energy more important than efficiency
 Power/Energy availability per volume/mass is important

Sources
 Motion
 Heat
 Light
 Radiation
 Stray fields
 Static charge

Energy harvesting is the collection of energy from the environment of a device, for local use

4 Speed vs Efficiency in Energy Flow

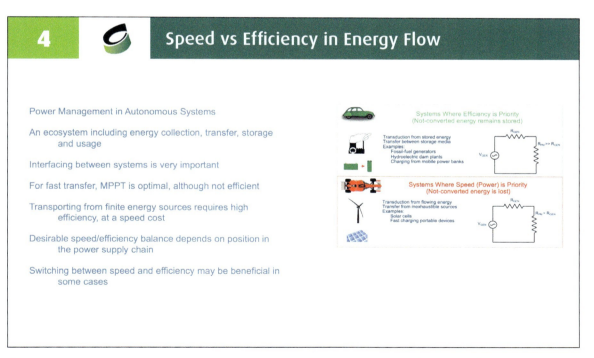

Power Management in Autonomous Systems

An ecosystem including energy collection, transfer, storage and usage

Interfacing between systems is very important

For fast transfer, MPPT is optimal, although not efficient

Transporting from finite energy sources requires high efficiency, at a speed cost

Desirable speed/efficiency balance depends on position in the power supply chain

Switching between speed and efficiency may be beneficial in some cases

Storage-to-storage power transfer requires maximum efficiency. In contrast, harvesting for a power-flowing source (e.g. sunlight) requires maximum power transfer. These are two different operation points.

5 Energy Harvesting as a Local Power Source (II)

Energy Harvesting is a Renewable Source
Sustainable and Environmentally Friendly
Not to power the grid

- Grid layer
  - 0.1 $/kWh
  - Cost is source dominated

- Portable power layer
  - 1000 $/kWh
  - Cost is storage dominated
  - Messy and unstructured

- The Internet of Things
  - 10000 $/kWh
  - Cost is Maintenance dominated

The "things" in the Internet of Things need power. This power is 5 orders of magnitude more expensive than grid-layer electrical power.

6 Energy Harvesting as a Local Power Source (III)

Harvesting is not for feeding the grid
The grid power is lower - cost
Portable power is very expensive
Battery replacement / recharging maintenance is even more expensive
Both economically and environmentally
Therefore, powering microsystems from local sources is very important

To summarize, using energy harvesting locally offers critical benefits in:
- Providing energy autonomy to wireless devices
- Making WSNs with more than 10 nodes practical
- Efficient energy distribution
- The environmental impact of the IoT

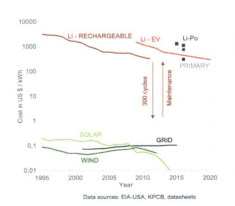

Energy harvesting can make WSNs with more than 10 nodes practical

7 Power Availability

Powering source	Power density State-of-Art	2025 (anticipated)
Direct thermoelectric	0.5 mW/cm² @ ΔT=5 K 10 mW/cm² @ ΔT=20 K	× 10 (superlattices)
Dynamic thermoelectric	1 mW/g in 2-hour flight 1 µW/g from day/night cycle	10 mW/g in 2 hour flight 1 mW/g from day/night cycle
Outdoor solar	10 mW/cm² @ direct sun 2 mW/cm² daily average	× 2 (dual bandgap stacks)
Indoor solar	10 µW/cm², diffused 400 lm	× 2 (dual bandgap stacks)
Airflow	4 mW/cm² @ 10 m/s	× 2 (MEMS Scaling)
Motion	0.1 mW/cm³, 100 Hz 10 µW/g, 700 Hz Eurocopter	Scaling and broadband
Inductive WPT	1 mW @ range = 5 × size	Longer range (directionality, rectennas) } Not Energy Harvesting!
Acoustic WPT	1 mW/cm³ @ 1 m of metal	× 10 (scaling, tuning)

Sources:
[1] Off-the-shelf TEG specs, e.g. Marlow, Eureca
[2] Kiziroglou et al, MST 2016
[3] For 10% efficiency, 5.5 hours per day
[4] Internal data, UC Berkeley, Aug. 2016
[5] Howey et al, SMS, 20, 085021, 2011
[6] Waterbury and Wright, JMES, 227, 1187, 2012
[7] Boyle et al, IEEE Perv. Computing, 15, 28, 2016
[8] Kiziroglou et al, PowerMEMS 2015

Power in the range of 1 mW/g is available in some environments. These numbers are at the transducer output, i.e. before power management / storage.

8 Summary of Thoughts

Energy harvesting is a special case of renewable energy technologies
Energy harvesting is not for feeding the grid
It is for powering small, local, wireless systems
The state-of-art can provide a few mW of power only

But we are in a CPS School, which involves embedded microsystems – GOOD!

Note the current challenge: Enough energy, but custom solution for each application required.
Possible next-decade stage: Power supplies that work in a specific environment
                              (e.g. in a car, around specific infrastructure, in an industrial plant,…)

The available power is enough for various duty cycled cyber physical systems. However, customization to each application is required. This is a significant limiting factor / cost-burden.

HETEROGENEOUS CYBER PHYSICAL SYSTEMS OF SYSTEMS

9. Motion Energy Harvesting (I)

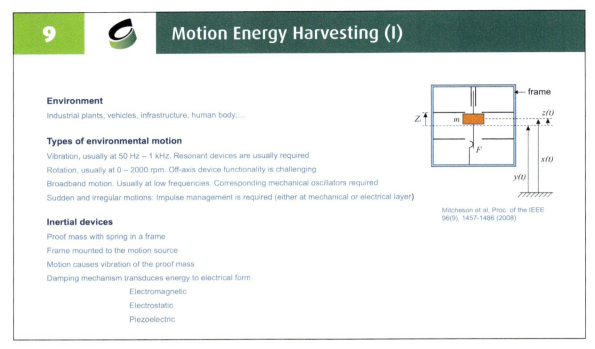

Environment
Industrial plants, vehicles, infrastructure, human body,...

Types of environmental motion
Vibration, usually at 50 Hz – 1 kHz. Resonant devices are usually required
Rotation, usually at 0 – 2000 rpm. Off-axis device functionality is challenging
Broadband motion. Usually at low frequencies. Corresponding mechanical oscillators required
Sudden and irregular motions: Impulse management is required (either at mechanical or electrical layer)

Inertial devices
Proof mass with spring in a frame
Frame mounted to the motion source
Motion causes vibration of the proof mass
Damping mechanism transduces energy to electrical form
- Electromagnetic
- Electrostatic
- Piezoelectric

Mitcheson et al, Proc. of the IEEE 96(9), 1457-1486 (2008)

Environmental motion is typically low frequency and broadband. This requires non-resonant transducers.

10. Motion Energy Harvesting (II)

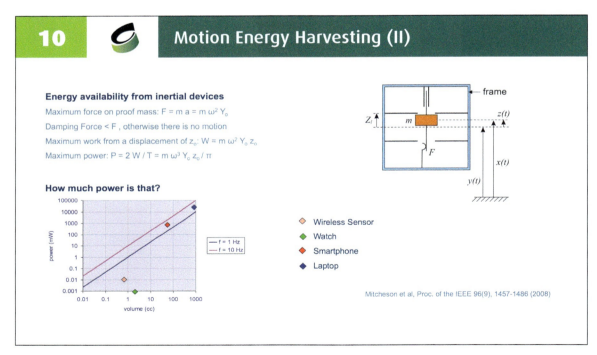

Energy availability from inertial devices
Maximum force on proof mass: $F = m a = m \omega^2 Y_0$
Damping Force < F, otherwise there is no motion
Maximum work from a displacement of z_0: $W = m \omega^2 Y_0 z_0$
Maximum power: $P = 2 W / T = m \omega^3 Y_0 z_0 / \pi$

How much power is that?

- ◇ Wireless Sensor
- ◆ Watch
- ◆ Smartphone
- ◆ Laptop

Mitcheson et al, Proc. of the IEEE 96(9), 1457-1486 (2008)

For a given vibration, the maximum power is $P = m \omega^3 Y_0 z_0 / \pi$. This for example is 1 mW/cm³ at 1 Hz.

11. Motion Energy Harvesting (III)

Challenges

Resonant operation → suitability only for very specific environments
Low-frequency, broadband and irregular motion management

$$f = \frac{1}{2\pi}\sqrt{\frac{k}{m}} \qquad k = \frac{E \cdot w \cdot \tau^3}{4 \cdot L^3}$$

E: Young's Modulus
L, w, τ: Beam Length, width and thickness
m: Mass

Frequency up-conversion

Oscillator that transforms motion to a higher frequency vibration

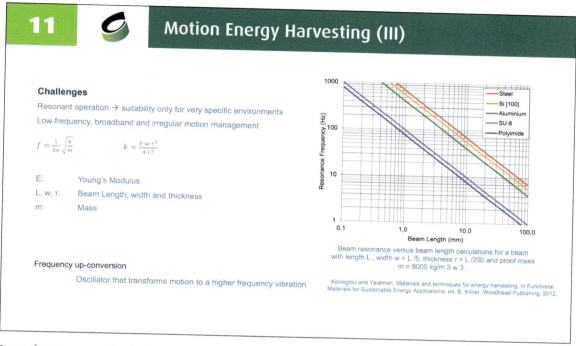

Beam resonance versus beam length calculations for a beam with length L, width w = L/5, thickness τ = L/200 and proof mass m = 8000 kg/m 3 w 3.

Kiziroglou and Yeatman, Materials and techniques for energy harvesting, In Functional Materials for Sustainable Energy Applications, ed. E. Kilner, Woodhead Publishing, 2012.

Low frequency mechanical resonators are difficult to make in the small scale.

12. Motion Energy Harvesting: Transduction Methods

Electrostatic

Moving plate of charged capacitor
The additional energy comes from the work done by the electrostatic force during plate displacement

$$V_{ouput} = \frac{C_{input}}{C_{output}} V_{input} \qquad \Delta E = \frac{1}{2} C_{output} V_{ouput}^2 - \frac{1}{2} C_{input} V_{input}^2 \qquad Q = C_{input} V_{input} \qquad Q = C_{output} V_{output}$$

Piezoelectric

Practically the same by at atomic level!
Strain changes interatomic distances, causing charge re-arrangement
Coupling between strain S and electric field E $\quad S = d \cdot E$
Most of energy is stored mechanically (i.e. no net E) but some can be exploited electrically

Electromagnetic

Faraday's law of induction $\qquad V = -\frac{d\Phi}{dt}$
Changing flux can originate from the field source but also from motion

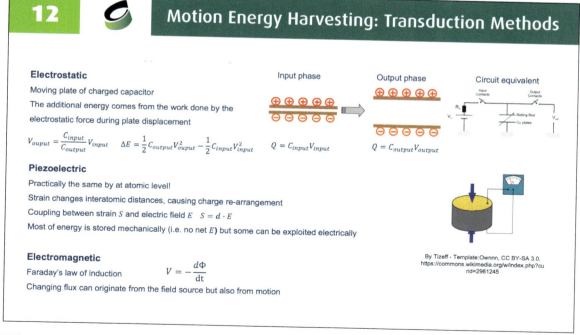

By Tizeff - Template:Ownnn, CC BY-SA 3.0, https://commons.wikimedia.org/w/index.php?curid=2961245

Electrostatic harvesters transduce motion energy into electricity through the work done by the electrostatic force during the displacement of a capacitor's plate.

13 — Motion Energy Harvesting: Electrostatic Example (I)

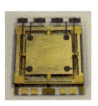
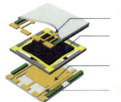

- Discharge contact on top plate
- Moving capacitor plate / mass
- Fixed capacitor plate on baseplate
- Pre-charging contact

Miao P., Mitcheson P.D., Holmes A.S., Yeatman E.M., Green T.C., Stark B.H., "MEMS inertial power generators for biomedical applications", Microsystem Technologies, 12, (2006), 1079-1083.

- Capacitor pre-charged when mass at bottom (max capacitance)
- Under sufficiently large frame acceleration, capacitor plates separate *at constant charge*, work is done against electrostatic force
- Charge transferred (at higher voltage) to external circuit when moving plate reaches top plate
- Non-resonant suspension: broadband
- A few µW at 1-10 Hz

A first MEMS prototype was introduced in 2016 by Miao et al., providing a few µW at 1-10 Hz.

14 — Motion Energy Harvesting: Electrostatic Example (II)

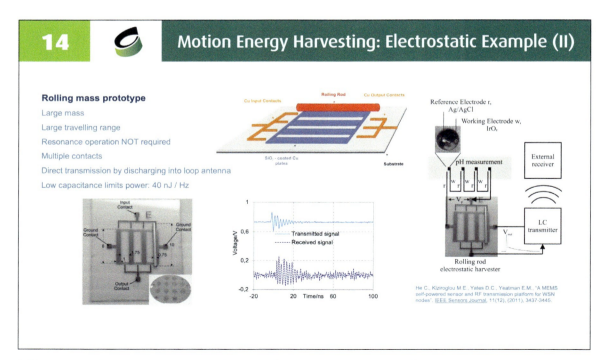

Rolling mass prototype
- Large mass
- Large travelling range
- Resonance operation NOT required
- Multiple contacts
- Direct transmission by discharging into loop antenna
- Low capacitance limits power: 40 nJ / Hz

He C., Kiziroglou M.E., Yates D.C., Yeatman E.M., "A MEMS self-powered sensor and RF transmission platform for WSN nodes", IEEE Sensors Journal, 11(12), (2011), 3437-3445.

Using external and free-moving proof masses allowed operation at lower frequency with promise for higher power density. The first self-powered wireless sensor was demonstrated in 2011.

The introduction of MEMS on flexible substrates allowed architectures designed for impulse excitation and mechanical frequency up-conversion, leading to 0.5 µW from human-like impulse motion.

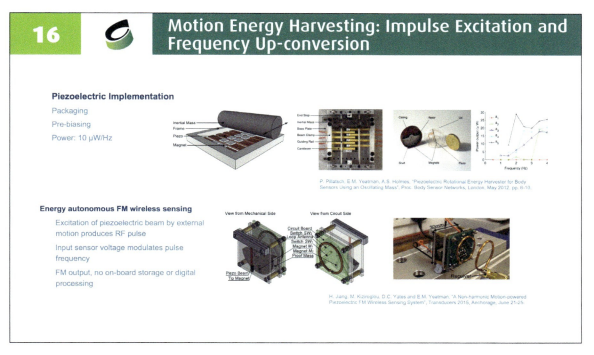

Impulse excitation or low frequency motion can be up-converted by magnetically plucking piezoelectric beams.

17 — A Rotational Motion Example: The FliteWISE system (I)

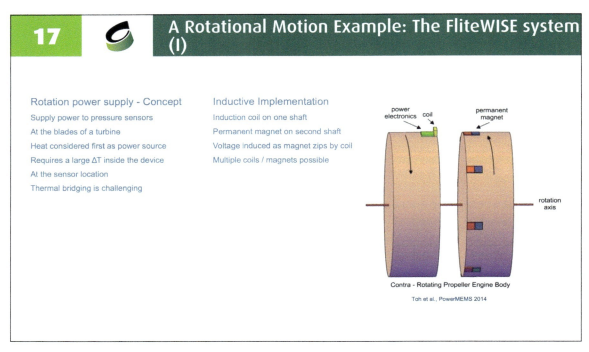

Rotation power supply - Concept
- Supply power to pressure sensors
- At the blades of a turbine
- Heat considered first as power source
- Requires a large ΔT inside the device
- At the sensor location
- Thermal bridging is challenging

Inductive Implementation
- Induction coil on one shaft
- Permanent magnet on second shaft
- Voltage induced as magnet zips by coil
- Multiple coils / magnets possible

Contra - Rotating Propeller Engine Body
Toh et al., PowerMEMS 2014

Inductive energy harvesting can be used in environments with steady relative motion between two neighbouring shafts (e.g. from rotation). A coil can pick up power from the changing flux of a passing magnet.

18 — A Rotational Motion Example: The FliteWISE system (II)

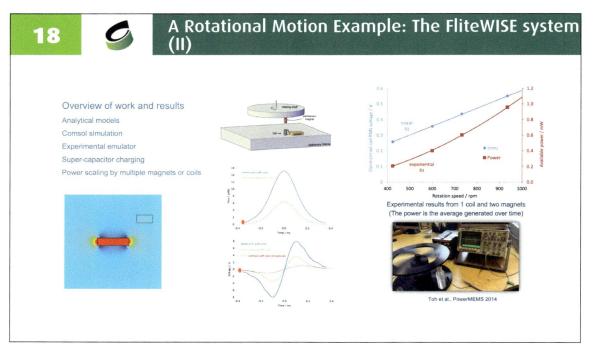

Overview of work and results
- Analytical models
- Comsol simulation
- Experimental emulator
- Super-capacitor charging
- Power scaling by multiple magnets or coils

Experimental results from 1 coil and two magnets
(The power is the average generated over time)
Toh et al., PowerMEMS 2014

Simulation can be used to predict the output power and optimize the design of inductive harvesters.

19 Thermal Energy Harvesting (I)

Typically using thermoelectric generators (TEGs)
- Based on the Seebeck effect in two materials
- ΔT across TEG builds up a voltage: $V = (S_A - S_B) \cdot \Delta T$
- The Seebeck coefficient S depends on DOS asymmetry at Fermi level
- S values are in the tens of µV/K range
- TEGs comprise arrays of couples, reaching tens of mV/K

Effect of current flow
- The Peltier effect (Inverse Seebeck effect)
- An Ohmic effect
- The Thomson effect

Thermoelectric generators transducer heat to electrical energy by the Seebeck effect.

20 Thermal Energy Harvesting (II)

Calculation of a TEG specifications from material properties
- N – couples, 100% volume filling
- Materials with S_A and S_B
- Thermal resistivity ρ_{TH}
- Electrical resistivity ρ_E
- TEG surface area A and thickness τ

Open circuit voltage: $V_{OC} = N \cdot (S_p - S_n) \cdot \Delta T$

Electrical resistance: $R_E = \rho_E \cdot \frac{2 \cdot N \cdot \tau}{A/2N} = \rho_E \cdot \frac{\tau}{A} \cdot 4 \cdot N^2$

Thermal resistance: $R_{TH} = \rho_{TH} \cdot \frac{\tau}{A}$

For a given ΔT, max output power is obtained on a load $R_L = R_E$

$$P_{out} = \frac{A}{16 \cdot \rho_E \cdot \tau}(S_p - S_n)^2 \cdot \Delta T^2 = \frac{V_{OC}^2}{4 \cdot R_E}$$

For small ΔT and ZT values, efficiency is a few %

The effects of current can be neglected, allowing: $P_{in} = \frac{\Delta T}{R_{TH}}$

Then, the efficiency is: $\eta = \frac{P_{out}}{P_{in}} = \frac{\rho_{TH}}{16 \rho_E}(S_p - S_n)^2 \cdot \Delta T = \frac{ZT}{16} \cdot \frac{\Delta T}{T}$

Notes:
This is the max power point, not max η
Here, ZT is at material level: $ZT = \frac{\rho_{TH}}{\rho_E}(S_p - S_n)^2 T$
The material ZT is 4 × ZT of device (using R_{TH} and R_E). (Due to geometry.)

Conclusion:
Efficiency depends on material and ΔT only
P_{OUT} scales with ΔT^2

The design of thermoelectric generators determine their voltage level, electrical resistance and thermal resistance, but does not affect the efficiency. The efficiency is determined by the material and the applied ΔT only.

21 — Thermal Energy Harvesting (III)

TEG power availability in practice
- Real environment: ΔT comes with a series thermal resistance $R_{CONTACTS}$
- TEG R_{TH} defines heat flow (incoming power)
- Large R_{TH} → high ΔT but low $\dot{Q}$. Small R_{TH} → high $\dot{Q}$ but low ΔT
- And low ΔT means low conversion efficiency
- Maximum electrical power at $R_{TH} = R_{CONTACTS}$
- Conclusion:
    - Minimization of $R_{CONTACTS}$ generally desirable
    - Once $R_{CONTACTS}$ is minimized, select R_{TH}
- TEG research front
    - Efficiency defined by material properties
    - Increase ρ_{TH}, reduce ρ_E.
    - Reduce phonon heat flow (Nanowires / Superlattices)
    - Increase S (e.g. thermionic emission)
    - Contacts can soon play a big role in thermoelectrics research

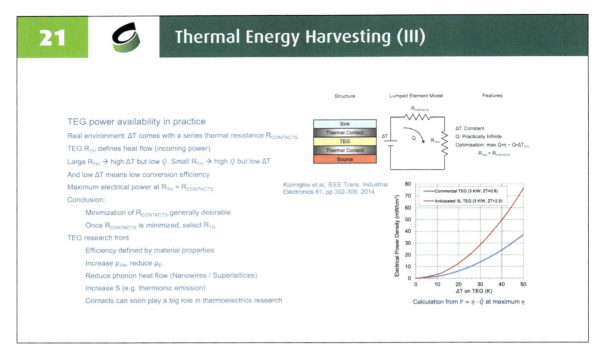

Kiziroglou et al, IEEE Trans. Industrial Electronics 61, pp.302-309, 2014

Calculation from $P = \eta \cdot \dot{Q}$ at maximum η

In practice, a TEG with thermal resistance that matches the thermal resistance of the contacts is needed for maximum power.

22 — Dynamic Thermoelectric Energy Harvesting (I)

Direct (Static) thermoelectric energy harvesting
- Requires a large ΔT inside the device
- At the sensor location
- Thermal bridging is challenging

Dynamic thermoelectric energy harvesting
- Also known as "heat storage thermoelectric harvesting"
- Exploitation of temperature variation in time
- Heat storage unit (HSU), thermally insulated from environment
- TEG between the HSU and the environment
- T_{IN} follows T_{OUT} with an RC exponential delay
- A PCM boosts heat storage

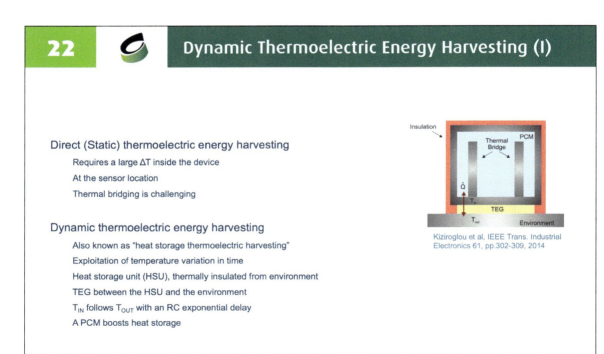

Kiziroglou et al, IEEE Trans. Industrial Electronics 61, pp.302-309, 2014

Dynamic heat storage can create an artificial **ΔT** from temperature fluctuations in time.

23 Dynamic Thermoelectric Energy Harvesting (II)

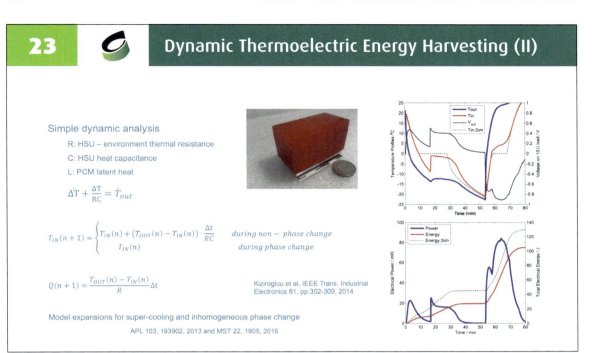

Simple dynamic analysis
R: HSU – environment thermal resistance
C: HSU heat capacitance
L: PCM latent heat

$$\dot{\Delta T} + \frac{\Delta T}{RC} = \dot{T}_{out}$$

$$T_{IN}(n+1) = \begin{cases} T_{IN}(n) + (T_{OUT}(n) - T_{IN}(n)) \cdot \frac{\Delta t}{RC} & \text{during non} - \text{phase change} \\ T_{IN}(n) & \text{during phase change} \end{cases}$$

$$Q(n+1) = \frac{T_{OUT}(n) - T_{IN}(n)}{R} \Delta t$$

Kiziroglou et al, IEEE Trans. Industrial Electronics 61, pp.302-309, 2014

Model expansions for super-cooling and inhomogeneous phase change
APL 103, 193902, 2013 and MST 22, 1905, 2016

Analytical and numerical models can accurately predict the performance and assist design optimization of dynamic thermal harvesters.

24 Dynamic Thermoelectric Energy Harvesting (III)

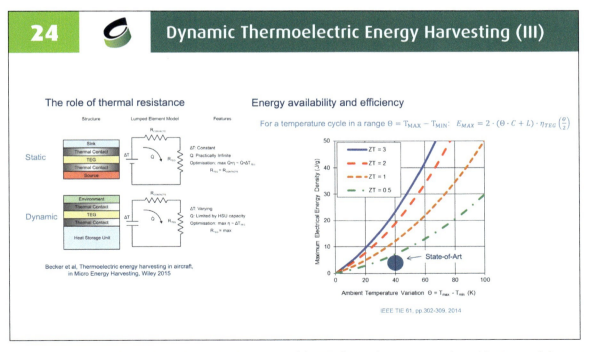

The role of thermal resistance

Energy availability and efficiency

For a temperature cycle in a range $\Theta = T_{MAX} - T_{MIN}$: $E_{MAX} = 2 \cdot (\Theta \cdot C + L) \cdot \eta_{TEG}\left(\frac{\Theta}{2}\right)$

Becker et al, Thermoelectric energy harvesting in aircraft, in Micro Energy Harvesting, Wiley 2015

IEEE TIE 61, pp.302-309, 2014

In dynamic thermoelectric harvesting the highest possible TEG thermal resistance is desirable. State-of-the-art of such devices deliver around 5 J/g per temperature cycle.

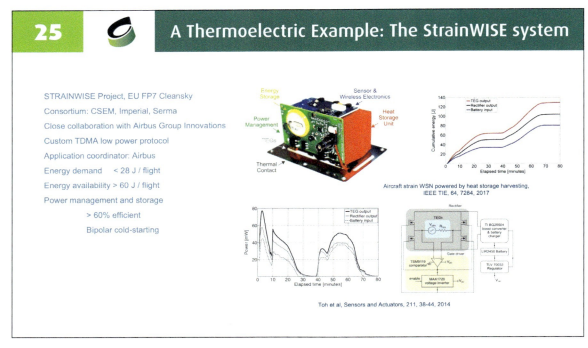

Dynamic thermoelectric harvesters can satisfy the power requirements of duty cycled aircraft strain wireless sensors.

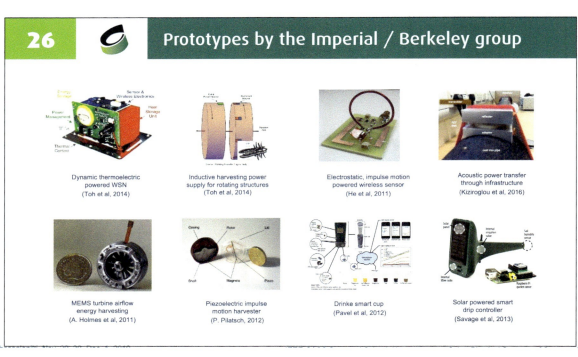

Our energy harvesting group has demonstrated a range of energy harvesting power supplies.

27 Summary

Objective: Energy autonomy of remote / mobile / wireless devices

Current status
- Energy harvesting can provide enough power
- Efficient management and storage available
- Low power electronics adequately low (1 mW active, 1 µW sleep, duty-cycling)
- But bespoke solutions required
- Cold starting is not generally available

Key desirable advancements (further to better materials)
- Near-zero sleep mode, near-zero leakage at storage
- Cold-starting by low voltage
- Energy harvesting from generally available conditions, e.g.
  - Around the human body
  - Around a vehicle
  - Across an infrastructure network
  - In a building

Energy harvesting is a very promising powering method for portable systems. The main challenge is currently its dependence to environmental conditions which limits applicability to bespoke solutions.

CHAPTER 08

Hands on Hardware / Software Co-Design

Dr. Nikolaos Tampouratzis

ECE School, Aristotle University
of Thessaloniki, Greece

1. Preliminaries (HDL to HLS) 256
2. Design Steps 256
3. Methodology 257
4. Chapter General Flow 258
5. Step 1. Project Creation 258
6. Design Under Consideration 259
7. Step 2. C-Simulation 260
8. Step 3. Design Synthesis 260
9. Step 4. Run C/RTL Co-Simulation 261
10. Step 5. Design Optimization (Loop Unrolling) 262
11. Step 5. Design Optimization (Loop Unrolling) 263
12. Step 5. Design Optimization (Loop Pipelining) 263
13. Step 5. Design Optimization (Loop Pipelining) 264
14. Step 5. Design Optimization (Arbitrary bit-widths) 265
15. Step 5. Design Optimization (BRAMs) 265
16. Step 6. Interface Declaration (SDSoC pragmas) 266
17. Step 6. Interface Declaration (SDSoC pragmas) 267

The breakdown of Dennard scaling coupled with the persistently growing transistor counts severally increased the importance of application-specific hardware acceleration; such approach offers significant performance and energy benefits compared to general-purpose solutions. The design of application-specific accelerators, until recently, has been predominantly done using Register Transfer Level (RTL) languages such as Verilog and VHDL, which, however, lead to a prohibitively long and costly design effort. In order to reduce the design time a wide range of both commercial and academic High-Level Synthesis (HLS) tools have emerged.

The strength of HLS is found in the ability to generate production quality Register Transfer Level (RTL) implementations from high-level specifications. In reality, this is a task that is constantly performed manually by design engineers and programmers while HLS promises to automate it, thus eliminating the source of many design errors and accelerating the currently very long development cycle.

Currently, one of the key challenges for the designer is to efficiently use the vendor-defined methodology and the design guidelines of the HLS tool. Hence, this chapter aims at assisting designers in taking full advantage and making optimal use of the Xilinx Vivado® HLS methodology when implementing a fundamental matrix multiplication algorithm which is used in a variety of application domains (machine learning, graphic, image, robotics, and signal processing applications). Finally, we use the Xilinx SDSoC™ environment in order to synthesize efficiently the proposed scheme to low-power Avnet Ultra 96 v1.2 board.

1 Preliminaries (HDL to HLS)

➢ Past → Hardware accelerators are mainly developed using Hardware Description Languages - HDLs
  ✓ Verilog, VHDL

➢ Today → Hardware accelerators can be developed using High-level programming languages through a number of tools (Cadence Stratus, **Xilinx Vivado HLS, etc.**)
  ✓ C, C++, SystemC

Advantages
♪ Reduce development time
♪ Quality of the design produced
♪ Manage and modify the original design easily → optimizations

Before we proceed in the lab, we will present an introduction about the trend of Hardware accelera-tor development. As we known, in the past Hardware accelerators are mainly developed using Hardware Description Languages – HDLs (Verilog, VHDL). However, today H/W accelerators can be developed using High-level programming languages adding an abstraction level in order to give the opportunity to S/W developers to design H/W through a number of tools (Cadence Stratus, Xilinx Vivado HLS, etc.) using languages such as C, C++, SystemC. Three of the most significant advantages are the following: (i) Development time reduction (orders of magnitude especially in the complex designs), (ii) Quality of the design produced and (iii) Management and modification of the original design applying a wide range of optimizations.

2 Design Steps

1. Design
2. Simulation
3. Implementation in Real H/W (e.g. board which includes FPGA SoC)

Designer develops an application (e.g. C/C++) which executed in the Processing System – **PS** (i.e. CPU) of FPGA SoC and **calls the accelerator** which is implemented in the Programmable Logic - **PL** (i.e. FPGA).

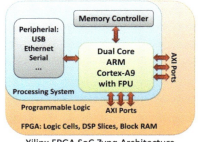

Xilinx FPGA SoC Zynq Architecture

2 Design Steps

After completing the design and simulation of a hardware accelerator, it is implemented in real hardware (e.g. a board which includes FPGA SoC) to check that the implemented accelerator adheres to the original design specifications when implemented in hardware. For this purpose, designer develops an application (e.g. C/C++) which executed in the Processing System – PS (i.e. CPU) of FPGA SoC and calls the accelerator which is implemented in the Programmable Logic - PL (i.e. FPGA). Figure illustrates the simplified architecture of Xilinx's FPGA SoC Zynq where the main parts of FPGA SoC, PS and PL are clearly visible.

3 Methodology

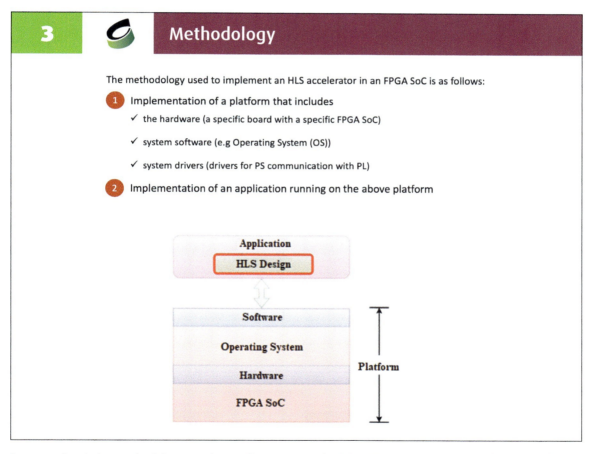

In more detail, the methodology used to implement an HLS accelerator in an FPGA SoC consists of 2 steps: (i) Implementation of a platform that includes hardware (a specific board with specific FPGA SoC) and system software (e.g. Operating System (OS)) and system drivers (drivers for PS communication with PL); (ii) Implementation of an application running on the above platform. Figure shows the procedure mentioned above as well as the "HLS Design" section which we will analyze in this chapter (red box).

4 Chapter General Flow

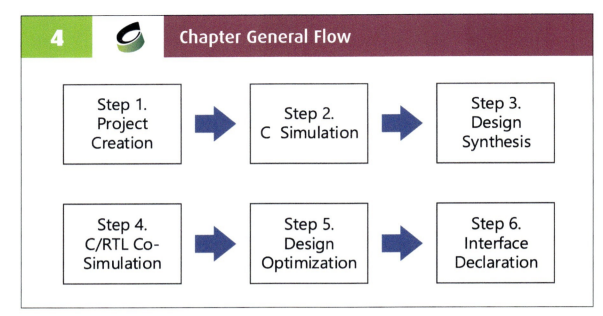

Specifically we use the Xilinx Vivado HLS tool for the implementation and optimization of our accelerator (application) as well as Xilinx SDSoC for the interface with the PS. This chapter is separated into steps that consist of general overview statements that provide information on the detailed instructions that follow. Especially, it comprises 8 primary steps; (i) Project creation, (ii) S/W-based Simulation, (iii) Design Synthesis, (iv) C/RTL Co-Simulation, (v) Design Optimization and (vi) Interface Declaration.

5 Step 1. Project Creation

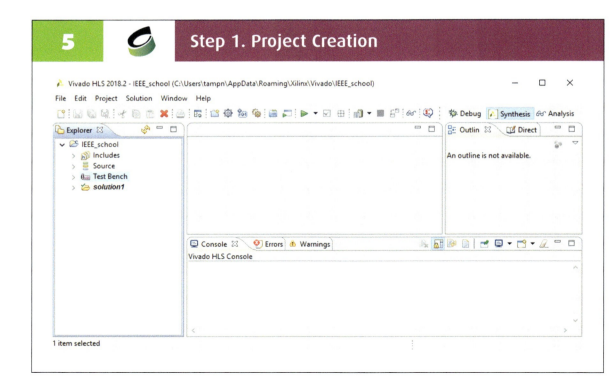

5 Step 1. Project Creation

During the project creation, Vivado HLS provides a simple wizard in order to specify the name of the project, the type of platform as well as the clock period. After the wizard, the user can observe a number of different things in the HLS window. First of all, the name of the project appears at the top of the Explorer window. In addition, Vivado HLS displays project information in a hierarchical way. Here we have information about the source code, the testbench file and the various solutions. The solution itself contains information about the implementation platform, design directives, and results (simulation, synthesis, IP export and more).

6 Design Under Consideration

$$(AB)_{ij} = \sum_{k=1}^{m} A_{ik} B_{kj}$$

Design (source code)   Testbench

It can be seen that the design is a matrix multiplication implementation, consisting of three nested loops. The Product loop is the inner most loop performing the actual Matrix elements product and sum. The Col loop is the outer-loop which feeds the next column element data with the passed row element data to the Product loop. Finally, Row is the outer-most loop. Line 19 resets the result every time a new row element is passed and new column element is used. In addition, a testbench has been implemented in order to verify the whole design executing the same matrix multiplication in s/w (right code). The TRIPCOUNT pragma can be applied to a loop to manually specify the total number of iterations performed by a loop in order to analyze the latency. To be noticed that the TRIPCOUNT pragma is for analysis only, and does not impact the results of synthesis.

7 Step 2. C-Simulation

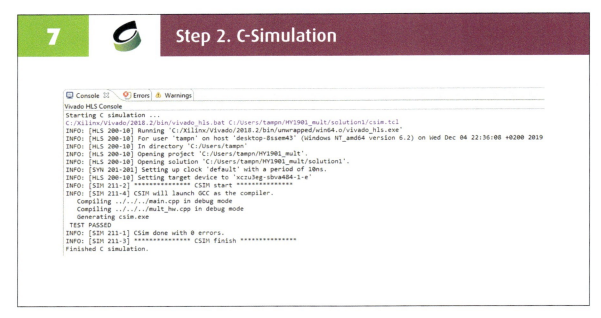

The next step in our design flow is to run a first simulation on this algorithmic level of design to verify that it performs the expected functionality. To do this, the user must press the Run C Simulation button or from the main menu Project > Run C Simulation. This option will first compile the code and then execute the design as a simple software program. Once the simulation has been completed successfully or failed, in the Console Panel can examine the printfs of the testbench to verify that the operations have been completed correctly, as well as examine any warning or error messages. If the user wish, in the Simulation Options window, he can select the Debug option and debugger to resolve any problems.

8 Step 3. Design Synthesis

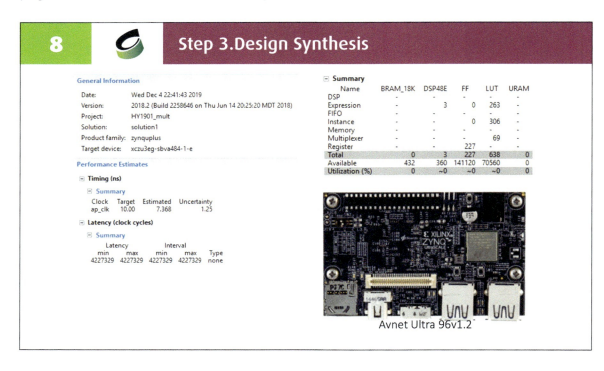

Avnet Ultra 96v1.2

8 Step 3. Design Synthesis

If the process completes successfully, the design is now ready for synthesis. In this step the tool converts the high-level description accelerator based on C ++ to lower-level description (RTL) VHDL, Verilog and SystemC languages (Run C Synthesis button). To be noticed that Avnet Ultra 96 v1.2 board is used. Here we can observe a number of things like: (i) the clock we set out to achieve (Estimated along with Uncertainty do not exceed the 10ns), (ii) the minimum and maximum latency (4.227.329 cycles). Also, the interval tells us when the next input is being read. In the minimum and maximum case, this is after 4.227.329 cycles of the clock. Therefore, our plan is NOT pipelined since a transaction must be completed before it can begin. Normally this is not the optimal behavior and we are modifying the functionality in the next part of this lab. In the right image, we can observe an estimation of Resources Utilization. Here we can get a first estimate of how many FPGA resources we have targeted to use. The most weighted fields are those of LUTs, FFs, DSPs and BRAMs. If the process completes successfully, the design is now ready for synthesis. In this step the tool converts the high-level description accelerator based on C ++ to lower-level description (RTL) VHDL, Verilog and SystemC languages (Run C Synthesis button). To be noticed that Avnet Ultra 96 v1.2 board is used. Here we can observe a number of things like: (i) the clock we set out to achieve (Estimated along with Uncertainty do not exceed the 10ns), (ii) the minimum and maximum latency (4.227.329 cycles). Also, the interval tells us when the next input is being read. In the minimum and maximum case, this is after 4.227.329 cycles of the clock. Therefore, our plan is NOT pipelined since a transaction must be completed before it can begin. Normally this is not the optimal behavior and we are modifying the functionality in the next part of this lab. In the right image, we can observe an estimation of Resources Utilization. Here we can get a first estimate of how many FPGA resources we have targeted to use. The most weighted fields are those of LUTs, FFs, DSPs and BRAMs.

9 Step 4. Run C/RTL Co-Simulation

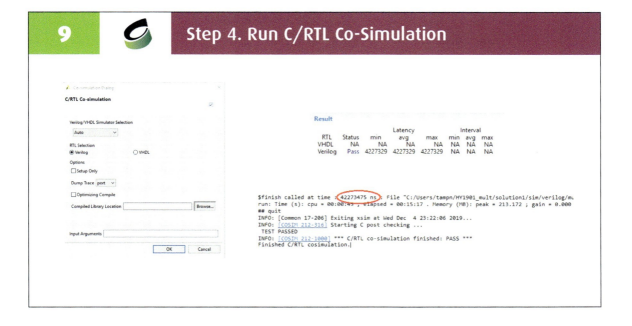

9 Step 4. Run C/RTL Co-Simulation

Now is the time to verify the RTL code that emerged through the synthesis stage. Here, our testbench file will be used to verify the RTL descriptions and the results of this process will be compared with the C-Simulation stage. When we click Run C/RTL Co-simulation from the toolbar, the above window will appear. Here our tool offers a number of immediate features such as specifying the HDL language (Verilog or VHDL RTL) in the verification process. Upon successful completion of this step, we are ready to examine the behavior of our system over time, according the testbench we set earlier, while in the right picture we can see the total execution time needed by our design (in cycles above & in time below).

10 Step 5. Design Optimization (Loop Unrolling)

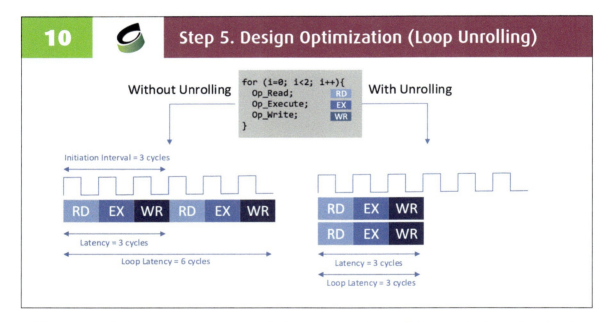

This step details how the user can control the hardware/architecture generated using directives in order to optimize it. Before we proceed in the Vivado HLS optimization techniques, we will describe some fundamentals about Unrolling and Pipeline. We use unroll loops to create multiple independent operations rather than a single collection of operations. We achieve that creating multiples copies of the loop body in the register transfer level (RTL) design, which allows some or all loop iterations to occur in parallel. Loops in the C/C++ functions are kept rolled by default. When loops are rolled, synthesis creates the logic for one iteration of the loop, and the RTL design executes this logic for each iteration of the loop in sequence; as a result the loop is executed for the number of iterations specified by the loop induction variable. Using UNROLLING, the user can unroll loops to increase data access and throughput.

11 — Step 5. Design Optimization (Loop Unrolling)

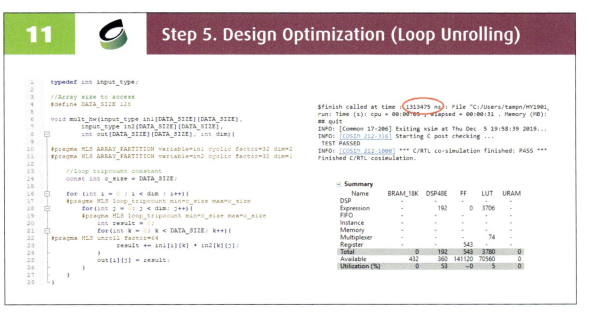

The UNROLL pragma allows the loop to be fully or partially unrolled. Fully unrolling the loop creates a copy of the loop body in the RTL for each loop iteration, so the entire loop can be run concurrently. Partially unrolling a loop lets you specify a factor N, to create N copies of the loop body and reduce the loop iterations. In the 1st optimization attempt we cannot use full unroll due to limited DSP resources (Ultra 96 is a low-end board with only 360 DSPs); as a result we use partially unrolled pragma by a factor of 64 in the most inner loop. Another serious limitation is the memory. Specifically, we would like to read 64 memory elements in one cycle, which is impossible since the memory can give us only up to 2 elements per cycle (the BRAM has up to 2 ports). For this reason, we use a process called array partitioning (lines 10 & 11) in order to divide the BRAM into smaller BRAM memories (32 is adequate because the code requires 64 concurrently accesses). After unroll optimization we achieve 32x speedup in performance, while the DSPs increases to 53% from 0%.

12 — Step 5. Design Optimization (Loop Pipelining)

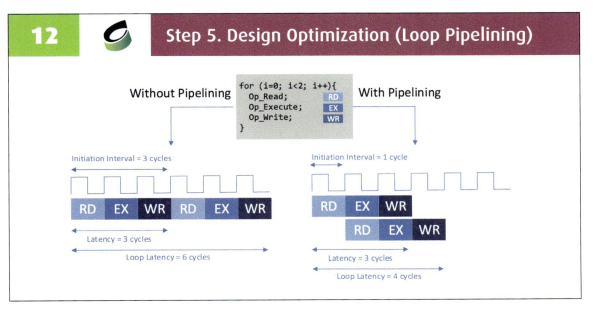

HETEROGENEOUS CYBER PHYSICAL SYSTEMS OF SYSTEMS

12 Step 5. Design Optimization (Loop Pipelining)

The PIPELINE method reduces the initiation interval for a function or loop by allowing the concurrent execution of operations. Pipelining a loop allows the operations of the loop to be implemented in a concurrent manner as shown in the figure. The left part of the figure shows the default sequential operation where there are 3 clock cycles between each input read (II=3), and it requires 6 clock cycles before the last output write is performed, while the right part shows the pipelined operation where there is only 1 cycle between each input read (II=1), and it requires 4 clock cycles before the last output write is performed. In other words, a pipelined loop can process new inputs every N clock cycles, where N is the initiation interval (II) of the loop.

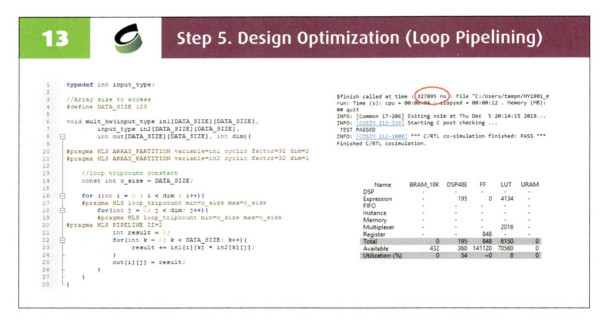

In the 2nd optimization attempt, we use pipeline pragma in order to achieve better performance. Due to limited DSP resources we use Initiation Interval (II) equal of 2. The main difference from the previous attempt is that (except from unrolling in k dimension) the pipeline pragma confirms that can process new 64 inputs in k dimension every 2 clock cycles (II=2). To be noticed that the HLS tool performs unroll in the code which is inside the pipeline pragma; it performs unroll with factor 64 because we set II=2 (the half of total k iterations). After pipeline optimization we achieve another 4x (128x total) speedup in performance (327.895ns instead of 1.313.475ns), while the DSPs remains almost constant.

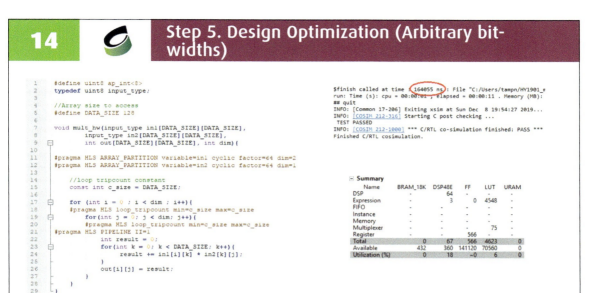

Step 5. Design Optimization (Arbitrary bit-widths)

The native data types in C are on 8-bit boundaries (8, 16, 32 and 64 bits). RTL signals and operations support arbitrary bit-lengths. Vivado HLS provides arbitrary precision data types for C to allow variables and operations in the C code to be specified with any arbitrary bit-widths (for example, 6-bit, 17-bit, and 234-bit, up to 1024 bits). The primary advantage of arbitrary precision data types is the Better quality hardware (e.g. when a 17-bit multiplier is required, the designer can use arbitrary precision types to require exactly 17 bits in the calculation). Without arbitrary precision data types, a multiplication such as 17 bits must be implemented using 32-bit integer data types. This results in the multiplication being implemented with multiple DSP48 components. In the 3rd optimization attempt, we use 8bit unsigned integer in order to achieve even better performance using II=1. After arbitrary bit-widths optimization we achieve another 2x (256x total) speedup in performance (164.055ns instead of 327.89ns), while the DSPs is decreased in half because a single 48bit DSP can execute 6 8bit operations instead of 1 32bit.

Step 5. Design Optimization (BRAMs)

Step 5. Design Optimization (BRAMs)

Up until this point, we have assumed that the data in arrays (in1 [][], in2[][], and out[][]) are accessible at anytime. In practice, however, the placement of the data plays a crucial role in the performance and resource usage. In most processor systems, the memory architecture is fixed and we can only adapt the program to attempt to best make use of the available memory hierarchy. In HLS designs, typically large amounts of data are stored in off-chip memory, such as DRAM, flash, or even network-attached storage. However, data access times are typically long, on the order of tens to hundreds (or more) of cycles. A common pattern is to load data into on-chip memory in a block, where it can then be operated on repeatedly. For this reason, in our example we need to copy the data into BRAMs so as to access a number of elements in one cycle (using ARRAY_PARTITION pragma). In the previous example this is not possible because the interface has not the required ports to give us concurrent elements in one cycle (BRAM utilization was 0%). After BRAM extension, the BRAM utilization becomes 29%, while we have a 2x penalty in performance.

Step 6. Interface Declaration (SDSoC pragmas)

```
#define uint8 ap_int<8>
typedef uint8 input_type;

//Array size to access
#define DATA_SIZE 128

void mult_hw(input_type* in1, input_type* in2, int* out, int dim){

    //loop tripcount constant
    const int c_size = DATA_SIZE;

    input_type BRAM_in1[DATA_SIZE][DATA_SIZE];
    input_type BRAM_in2[DATA_SIZE][DATA_SIZE];

    #pragma HLS ARRAY_PARTITION variable=BRAM_in1 cyclic factor=64 dim=2
    #pragma HLS ARRAY_PARTITION variable=BRAM_in2 cyclic factor=64 dim=1

    for (int i = 0 ; i < dim ; i++){
        #pragma HLS loop_tripcount min=c_size max=c_size
        for(int j = 0; j < dim; j++){
            #pragma HLS loop_tripcount min=c_size max=c_size
            #pragma HLS PIPELINE II=1
            BRAM_in1[i][j] = in1[i*DATA_SIZE + j];
            BRAM_in2[i][j] = in2[i*DATA_SIZE + j];
        }
    }

    for (int i = 0 ; i < dim ; i++){
        #pragma HLS loop_tripcount min=c_size max=c_size
        for(int j = 0; j < dim; j++){
            #pragma HLS loop_tripcount min=c_size max=c_size
            #pragma HLS PIPELINE II=1
            int result = 0;
            for(int k = 0; k < DATA_SIZE; k++){
                result += BRAM_in1[i][k] * BRAM_in2[k][j];
            }
            out[i][j] = result;
        }
    }
}
```

In the last step of our lab we use the Xilinx SDSoC™ environment which provides a framework for developing and delivering hardware accelerated embedded processor applications using standard programming languages. It includes compilers for the embedded processor application and for hardware functions implemented on the programmable logic resources of the Xilinx® device. The sds++ system compiler generates hardware IP and software control code that automatically implements data transfers and synchronizes hardware accelerators with the application software by invoking the Vivado HLS tool. In addition, the sds++ linker analyzes program dataflow involving calls into and between hardware functions, mapping into a system hardware data motion network, and software control code to orchestrate accelerators and data transfers. Data transfers between memory and accelerators are accomplished through data movers, such as a DMA engine, automatically inserted into the system by the sds++ system compiler taking into account user data mover pragmas. In our example, we transform the input array dimension to 1d in order to be accessed sequentially, while we use a number of SDS pragmas to achieve efficiently data transfers.

Step 6. Interface Declaration (SDSoC pragmas)

```c
#ifdef __SDSCC__
    #include <stdlib.h>
    #include "sds_lib.h"
    #define malloc(x) (sds_alloc(x))
    #define free(x) (sds_free(x))
#endif

#pragma SDS data sys_port( in1: AFI )
#pragma SDS data sys_port( in2: AFI )
#pragma SDS data sys_port( out: AFI )
#pragma SDS data copy( in1[0: DATA_SIZE*DATA_SIZE ] )
#pragma SDS data copy( in2[0: DATA_SIZE*DATA_SIZE ] )
#pragma SDS data copy( out[0: DATA_SIZE*DATA_SIZE ] )
#pragma SDS data data_mover( in1: AXIDMA_SIMPLE )
#pragma SDS data data_mover( in2: AXIDMA_SIMPLE )
#pragma SDS data data_mover( out: AXIDMA_SIMPLE )
#pragma SDS data access_pattern( in1: SEQUENTIAL )
#pragma SDS data access_pattern( in2: SEQUENTIAL )
#pragma SDS data access_pattern( out: SEQUENTIAL )
#pragma SDS data mem_attribute( in1: PHYSICAL_CONTIGUOUS )
#pragma SDS data mem_attribute( in2: PHYSICAL_CONTIGUOUS )
#pragma SDS data mem_attribute( out: PHYSICAL_CONTIGUOUS )
```

```
root@xilinx-ultra96-reva-2018_2:/home/app# ./demo.elf
Time duration in software:0.0142817
Time duration in hardware:0.000371177
Speed up: 38.4767
Note: Speed up is meaningful for real hardware execution only, not for emulation.
TEST PASSED
```

First of all, we use sds_alloc instead of simple malloc in order to allocate memory that is going to be physically contiguous taking advantage of some additional performance in communicating with the PL. In addition, #pragma SDS data sys_port(ArrayName:port) overrides the SDSoC compiler default choice of memory port. Specifically, the port must be either ACP/HPC (cache coherent interface) or AFI (non-cache coherent interface). In our example a simple Avnet Ultra 96 board is used which provides only AFI port. Furthermore, the 2nd pragma (#pragma SDS data copy) implies that data is explicitly copied between the host processor memory and the hardware function using a suitable data mover which performs the data transfer (in this case a simple DMA is used). Moreover the 4rd pragma specifies the data access pattern in the hardware function. If the access pattern is SEQUENTIAL, a streaming interface will be generated. Otherwise, with RANDOM access pattern, a RAM interface will be generated. Finally, with the 5th pragma we tell to SDSoC runtime to allocate physically contiguous memory. Summarizing, our implementation in hardware is faster two orders of magnitude (38x speedup) comparing with a low power ARM Cortex-A53 CPU.

About the Editors

Ioannis Papaefstathiou

Dr. Ioannis Papaefstathiou is an Associate Professor at the School of Electrical and Computer Engineering at Aristotle University of Thessaloniki and a co-founder of Exascale Performance Systems (EXAPSYS) which is a spin-off of Technical University of Crete, Synelixis Solutions SA and Foundation of Research and Technology Hellas (FORTH). From 2004-2018 he was a Professor at ECE School at Technical University of Crete and a Manager at Synelixis Solutions SA. His main research interests are in the design and implementation methodologies for CPS with tightly coupled design parameters and highly constrained resources as well as in heterogeneous High Performance Computing (HPC) systems and the associated programming/development tools. He was granted a PhD in computer science at the University of Cambridge in 2001, an M.Sc. (Ranked 1st) from Harvard University in 1997 and a B.Sc. (Ranked 2nd) from the University of Crete in 1996. He has published more than 100 papers in IEEE and ACM-sponsored journals and conferences, he has been the guest editor an IEEE Micro issues and he is a member of the Program Committee of numerous IEEE and ACM conferences. He is participating / has participated in numerous European R&D Programmes(e.g. OSMOSIS, FASTCUDA, HEAP, FASTER, COSSIM, ECOSCALE, EXTRA, EuroEXA); in total he has been Principal Investigator in 12 competitively funded research projects in Europe (in 7 of them he was the technical coordinator of the whole project), in the last 7 years, where his cumulative budget share exceeds €5 million.

Alkis Hatzopoulos

Prof. Alkis A. Hatzopoulos (Chatzopoulos) was born in Thessaloniki, Greece.

He received his Degree in Physics (with honours), his Master Degree in Electronics and his Ph.D. Degree in Electrical Engineering from the Aristotle University of Thessaloniki, Greece, in 1980, 1983 and 1989, respectively. He has been with the Department of Electrical and Computer Engineering at the Aristotle University of Thessaloniki since 1981, were now he is a full Professor. Since 2002 he has been elected as the Director of the Electronics Laboratory of the ECE Dept.

His research interests include modelling, design and testing of integrated circuits and systems (analog, mixed-signal, high-frequency), three dimensional Integrated Circuits (3D ICs), electronic communication circuits, thin-film transistors, instrumentation electronics, Bioelectronics systems, space electronics.

He is actively involved in educational and research projects, and he is the author or co-author of more than 170 scientific papers in international journals and conference proceedings and three book chapters in international textbooks. He has been also granted a European and American patent.

He has been a Member (2010-2018) of the Belgian Research Foundation Committee (FWO Expert panel WT7) for National (Flanders) Grants and Project evaluations. He has given a large number of lectures on his research topics in various Universities. He has been a visiting Professor at Michigan State University, MI, USA, (Feb. - July 1995), at the Katholieke Universiteit Leuven, Belgium (Feb. - July 2004), and at the University of Texas at Dallas (UTD) (Jan. - July 2016). He is a Senior Member of IEEE and he has been elected as the IEEE Greece CASS-SSCS joint Chapter Chair for many years.

Address:
Aristotle University of Thessaloniki
Dept. of Electrical and Computer Engineering
Electronics Lab.
Thessaloniki 54124, Greece
Tel.: +30 2310 996305
Tel.: +30 2310 996221 (Secretary)
Fax : +30 2310 996447
e-mail : alkis@eng.auth.gr   or : alkis@ece.auth.gr
url: http://ee.auth.gr/en/school/faculty-staff/electronics-computers-department/hatzopoulos-alkiviadis/
https://orcid.org/0000-0002-4030-8355

About the Authors

Eduardo de la Torre

Prof. Eduardo de la Torre has experience in adaptive systems and evolvable hardware. These technologies are applied in the context of embedded systems and cyber-physical systems. In the last years he is working in the area of scalable multithread hardware accelerators. He has participated in more than ten EU funded projects (two of them as IP for UPM), as well as in numerous projects for the National R&D Plans (two as IP) and also with projects funded by industry.

He has numerous publications in Journals and international conferences related with his research topics. He has been General Chair of DASIP 2014 and ReCoSoC 2017, and Program Chair of the Conferences Reconfig 2012, Reconfig 2013, Spie Microelectronics 2011, DASIP 2013 and DCIS 2014. He participates in Program Committees of the above mentioned conferences, as well as in ISVLSI, JCRA, RAW (IPDPS WorkShop), EUC, FPL or DATE. He has also been Special Issue Editor and/or reviewer of Journals such as IEEE Trans on Industrial Electronics, Journal of Systems Architecture, Journal in Advances in Signal Processing, MDPI Sensors, Electronics, IEEE Access, among others.

Apostolos Dollas

Prof. Apostolos Dollas received his Ph.D. in CS from the University of Illinois at Urbana Champaign (1987). He is currently Full Professor of ECE at the Technical University of Crete (TUC) in Chania, Greece, a Senior Research Associate of the CARV group of the Institute of Computer Science of FORTH (Irakleio), and a board member of the Telecommunication Systems Institute (Chania). At TUC, A. Dollas has served as Dean of the School of ECE (2013-2017), Director of the Microprocessor and Hardware Lab (1994-2009), and ECE Dept. Chairman (1995-1997). He has been Assistant Professor of ECE and CS at Duke University (1986-1994). He conducts research and teaches in reconfigurable computing, embedded systems, and application-specific high-performance computer architectures, with emphasis on fully functional prototypes. Dollas is a member of HKN and TBΠ honorary societies and has received the IEEE Computer Society Golden Core Member Award, and the IEEE Meritorious Service Award. He is co-founder of several international conferences, including FCCM, FPT, RSP, SASP, and TAI, and serves in several international conference program committees, including FPL (2011 General co-Chair); he is co-inventor in two issued US Patents.

Georges Gielen

Prof. Georges G.E. Gielen received the MSc and PhD degrees in Electrical Engineering from the Katholieke Universiteit Leuven (KU Leuven), Belgium, in 1986 and 1990, respectively. He is Full Professor in the MICAS research division at the Department of Electrical Engineering (ESAT). From August 2013 till July 2017 he was also appointed at KU Leuven as Vice-Rector for the Group of Sciences, Engineering and Technology, and he was also responsible for academic Human Resource Management.

His research interests are in the design of analog and mixed-signal integrated circuits, and especially in analog and mixed-signal CAD tools and design automation. He is a frequently invited speaker/lecturer and coordinator/partner of several (industrial) research projects in this area, including several European projects. He has (co-)authored 10 books and more than 600 papers in edited books, international journals and conference proceedings. He is Fellow of the IEEE since 2002, and received the IEEE CAS Mac Van Valkenburg career award in 2015. He is a 1997 Laureate of the Belgian Royal Academy of Sciences, Literature and Arts in the discipline of Engineering.

Ioannis Karamitsos

Dr. Ioannis Karamitsos received his PhD in Computer Science from University of Sunderland, UK, the Master degree in Telematics Management from University of Danube Krems, Austria, and BSc degree (Laurea) in Electronic Engineering from University of Rome "La Sapienza", Italy. He has vast industry and research experience over 25 years as an executive manager who worked within the private and public sectors and experienced within European, Middle East and Chinese companies. In 2016, he had joined the Department of Electrical Engineering as a faculty member at Rochester Institute of Technology Dubai. Dr. Ioannis is particularly interested in applying Blockchain, Cryptography, Machine learning, IIoT, and data mining techniques to emerging problems related to large-scale decentralized cyber-physical systems and critical infrastructures as well as energy, mining, health care and other domains of major economic and social impact.

Michail E. Kiziroglou

Dr. Michail E. Kiziroglou obtained his diploma in electrical and computer engineering from Aristotle University of Thessaloniki, Greece, in 2000 and his master in microelectronics from Democritus University of Thrace, Greece, in 2003. He holds a Ph.D. in microelectronics and Si spintronics awarded by the University of Southampton in 2007. Between 2006 and 2018 he has been a research associate with the optical and semiconductor devices group, department of electrical and electronic engineering, Imperial College London. He is currently an assistant professor at the department of industrial engineering and management, International Hellenic University, Greece, and a research fellow at Imperial College London. In 2016 he worked as an associate project scientist at the department of mechanical engineering, University of California at Berkeley, on the development of microgenerators for aircraft applications. He has over 60 publications in international journals and conferences, 30 of which on energy harvesting devices. He is a senior member of the IEEE and a member of the Institute of Physics. His research interests include energy harvesting devices, microengineering and energy autonomous wireless sensors.

Harry Manifavas

Dr. Harry Manifavas holds the position of Senior Research Collaborator at ICS-FORTH. His experience and expertise lie in the areas of Cryptography, Network and Information Systems Security. His current areas of interest include mobile forensics, embedded systems security, cybersecurity strategies and blockchain. He received his Ph.D. in Computer Science from the University of Cambridge (UK), his M.Sc. in Communication Systems Engineering from the University of KENT (UK), and his B.Sc. in Computer Science from the University of Crete (Greece). From 1995 to 1998 he worked as a Research Assistant at the Computer Lab, University of Cambridge, on electronic payment protocols, digital certification infrastructures and copyright protection. From 2000 to 2003 he worked as a security engineer for Barclays Capital, an investment bank in London. From 2007 to 2010 he worked as a security engineer & consultant for Virtual Trip, Ltd., FORTH-ICS and ENISA. From 2011 to present he has participated as technical coordinator or senior security engineer in a number of EU funded research projects and served twice as an evaluator of EU research proposals. From 2004 to present, he has been lecturing at several universities on Information Systems Security and Network Security. During this time, he supervised several MSc and PhD theses. He is the author or co-author of more than 50 research papers and a book, all in the area of information security. He has received professional certifications by ComTIA, IRCA, Terena and MSAB.

Nikolaos Tampouratzis

Dr. Nikolaos Tampouratzis is a computer scientist and researcher at School of Electrical and Computer Engineering, at Aristotle University of Thessaloniki (AUTH) in Greece. He received his BSc from University of Crete, and MSc and Ph.D from Technical University of Crete. He has more than 10 publications in international journals and conferences on embedded/hardware design and simulators. He has been involved for more than 8 years as a researcher in numerous European and National competing research programs towards the design, development and validation of state-of-the-art technologies in embedded/hardware design and simulators. Currently, apart from AUTH, he also provides research and development services to EXAPSYS.

Theocharis Theocharidis

Prof. Theocharis Theocharides has been an Associate Professor at the Department of Electrical and Computer Engineering, University of Cyprus, Nicosia, Cyprus, since 2006, where he directs the Embedded and Application- Specific Systems-on-Chip Laboratory. He has also been a Faculty Member of the KIOS Research and Innovation Center of Excellence, University of Cyprus, since the Center's inception in 2008. His research focuses on the design, development, implementation, and deployment of low-power and reliable on-chip application-specific architectures, low-power VLSI design, real-time embedded systems design, and exploration of energy-reliability tradeoffs for systems on chip and embedded systems. His focus lies on acceleration of computer vision and artificial intelligence algorithms in hardware, geared toward edge computing, and in utilizing reconfigurable hardware toward self-aware, evolvable edge computing systems. Theo has a PhD in computer engineering from Penn State University, State College, PA, working in the areas of low-power computer architectures and reliable system design with emphasis on computer vision and machine learning applications. He is a Senior Member of the IEEE and a member of CEDA and the ACM. He is currently serving in the editorial boards of the IEEE Design & Test Magazine, and is an Associate Editor of ACM Journal of Emerging Technologies and IEEE Consumer Electronics Magazine, as well as IET Computers and Digital Techniques and the ETRI Journal.

Anna-Maria Velentza

Anna-Maria Velentza received her BSc degree in Psychology from University of Crete, GR in 2015 and her MSc degree from University of Birmingham, UK, in 2018 in Computational Neuroscience and Cognitive Robotics.

She is currently a PhD Student in the Laboratory of Informatics and Robotics for Education and Society (LIRES), University of Macedonia, Greece, and her research mainly focus in the role of cognitive functions such as memory and attention in Human Robot Interaction.

She has worked in several research institutes in multidisciplinary projects in cognitive psychology and robotics (e.g. Model Comparison on perceptual decision making as part of the center for neuroscience and robotics of the University of Birmingham, Birmingham, UK, study of existing systems and models for human-robot interaction as part of the Telecommunication System Institute, Technical University of Crete). From 2014 to 2017 she was a member of the Transformable Intelligent Environment Laboratory, Technical University of Crete, Chania, Greece designing and carrying out cognitive experiments on the effects of environmental conditions on perception, attention and mood, while for 2 years she was the Coordinator of the Psychospatial Department, a multidisciplinary student group of architects, mechanical engineers and electrical engineers. Her current research interests are on the field of human- robot interaction and human-robot collaboration mainly focusing in novel methods to enhance robots' cognitive functions.